Risk-Based Analysis for Environmental Managers

Edited by
Kurt A. Frantzen

CRC Press
Taylor & Francis Group
Boca Raton London New York

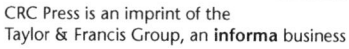

CRC Press is an imprint of the
Taylor & Francis Group, an **informa** business

CRC Press
Taylor & Francis Group
6000 Broken Sound Parkway NW, Suite 300
Boca Raton, FL 33487-2742

First issued in paperback 2019

© 2002 by Taylor & Francis Group, LLC
CRC Press is an imprint of Taylor & Francis Group, an Informa business

No claim to original U.S. Government works

ISBN-13: 978-1-56670-379-6 (hbk)
ISBN-13: 978-0-367-39676-3 (pbk)

Visit the Taylor & Francis Web site at
http://www.taylorandfrancis.com

and the CRC Press Web site at
http://www.crcpress.com

Risk-Based Analysis for Environmental Managers

PREFACE

Early in my career, I had the opportunity to perform a risk assessment for a state Superfund site in Erie County, New York. It just so happened that a friend and colleague of mine lived nearby. At the time, Dave was working on his doctoral degree in toxicology. Even though we were similarly trained, both educated in scientific disciplines, and friends, when we spoke about the site his attitude was quite different from mine; he expressed genuine concern over the situation, whether the site was going to be cleaned up, and whether or not we really knew what was going on. Although pleasant about the discussion, it never seemed possible to allay his fears.

In 1998, after working together on a project, John Voorhees, an environmental lawyer and co-author of *International Environmental Risk Management* (1998), suggested that I write a book about risk assessment as part of the environmental engineering/risk management series of the Lewis Publishers group of CRC Press. After reflecting on the idea, the proposition did not seem necessary because of the many good books already on the market addressing the subject. Instead, observing the frustration of those who use the products of risk assessment or live with the results of them, I reckoned that perhaps this audience needed a book to help them see the issue of environmental risk and the role of risk assessment from a large, let us say "macro" view, and how to use the tool and manage risk assessors better in their own projects.

Thus, this book seeks to be a *practical guide*, and it is primarily for corporate and public service management personnel assigned to environmental issues. Its particular focus is on how to manage the assessment of health and ecological risks associated with "impaired" real estate. Using this guide, such personnel will be better able to control the definition of risk and influence the environmental decision-making process, including environmental negotiations with regulators, other stakeholders, and the public. We call the approach "Risk-Based Analysis" (RBA); it is generic and applicable to the array of regulatory schemes now extant (in the United States and many in European countries). The approach focuses on the problem of the definition of risk, which arises at the interface of risk assessment, risk management, and risk communication. This definition (and its qualification and quantification) drives the legal, financial, property value, and image issues bedeviling many owners of environmentally impaired property. Through a simple, five-step technique, Risk-Based Analysis serves as a framework for the management of environmental risk that will not only ensure protection of human health and the environment, but also that the asset value of the impaired property and the associated liability are effectively managed, with image and credibility enhanced, and community viability and socioeconomic standing augmented. The book contains background information describing the RBA concept and its development, management tools and techniques, and perspective on the role of the risk issue in financial decision-making related to contaminated properties in the urban-industrial landscape.

As I was writing this preface, I received the latest copy of *Risk Analysis*. In it was a perspective article written by Yacov Haimes titled "Risk Analysis, Systems

Analysis, and Covey's Seven Habits" (21(2) 217 – 224, April 2001). In the article, Haimes discusses the commonality of systems analysis and risk analysis in using a holistic/gestalt approach to problem solving, and he shows that Covey's elements are similarly present and complementary in both. He encourages practitioners (and I would hasten to add those who use the products of those practitioners) to keep in mind those elements as they work on the complicated problems spanning our large-scale and technically complex world and the societal systems therein. To achieve success in resolving problems associated with environmental risk management issues of impaired properties, it is important to consider all of the relevant dimensions and perspectives. In doing so, and beginning with the end in mind (as Covey says), Risk-Based Analysis will help Environmental Managers produce not expenses, but revenues; not obstacles, but opportunities; not an imposition, but a better quality of life.

EDITOR

Kurt A. Frantzen is currently managing director of Environment Risk Management at Vanasse Hangen Brustlin, Inc. (VHB). He specializes in defining and solving technical and perceptual issues arising out of liability concerns associated with environmental contamination, property impairment, and suspected human and ecological exposures. Dr. Frantzen has provided expert technical support for field activities, developed methodologies to assess health and ecological risk, and applied the results to influence decision-making to obtain practical, cost-effective solutions. He has provided agency liaison and negotiated priorities for regulatory compliance. As a consultant and communicator working in private, public, administrative, and legal forums, Dr. Frantzen seeks to facilitate client and stakeholder understanding of and allay fears associated with complex technical issues and uncertainties surrounding environmental impairment. His clients include major industrial corporations, gas and electric utilities, law firms, federal and state agencies, and institutional organizations. Dr. Frantzen has published articles and spoken widely on environmental risk analysis and management to professional associations and colleges. He received his Ph.D. in biochemistry from the University of Nebraska at Lincoln in 1985, an M.S. degree in plant pathology from Kansas State University in 1980, and a B.S. degree in biology from the University of Nebraska at Omaha in 1978. He also was an American Cancer Society Post-Doctoral Fellow at the University of Washington in Seattle. He is a member of the American Chemical Society, American Institute of Biological Sciences, and the Society for Risk Analysis.

CONTRIBUTORS

Jerry Ackerman is a marketing consultant and strategic planning analyst for commercial, industrial, and governmental clients engaged in the management of environmentally impaired properties. His management consulting background in addressing site remediation and redevelopment issues includes: developing business plans, structuring technical and financial resource partnerships; consensus building; marketing redevelopment programs; and conducting Brownfields-related educational workshops. He and Dr. Frantzen have collaborated on the design and development of communications tools, land-use proposals, educational materials, and business decision alternatives. With degrees in English and socioeconomics, Mr. Ackerman has published nearly 100 articles on the sustainable, economic redevelopment of Brownfields and other environmentally impaired properties, and he has spoken widely on the subject throughout the United States and Europe.

Samuel D. Ostrow founded Ostrow & Partners in 1994, after twenty-five years with major public relations consultancies, including serving as head of the public affairs and crisis management practices of two of the ten largest firms. He has developed corporate public affairs programs for many Fortune 100 companies, and worked in issues and the crisis management area developing strategies and directing programs for several pharmaceutical and financial services. He has worked with Brooklyn Union Gas, Samarec, McDonald's, Sandoz, Tambrands, the Aluminum Association, Crown Butte, and Hudson Cedar on environmental issues, and LILCO on nuclear power plant emergency plans. He also has worked on U.S.-Japan trade issues, food safety issues, and on airline safety. He has directed communications in over 60 corporate mergers and acquisitions, representing acquirers as well as targets, and has developed and directed investor relations programs for companies ranging from the Dow 30 to start-ups in the computer and biotechnology fields. Mr. Ostrow is a frequent writer and speaker on public relations topics. He is past president of the Chicago Chapter of the National Investor Relations Institute (NIRI); a former director of the New York Chapter of NIRI; and a director of Global Liability Management SA, the international crisis management consortium. He is president of the Pound Ridge Association, a founding member of the Citizens' Committee for the Hiram Halle Library, a member of the Dartmouth College Alumni Council, President of his Dartmouth College class, a Director of the Stamford Boys & Girls Club, and an honors graduate of The John Marshall Law School.

John Rosengard is President of Environmental Risk Communications, Inc. (ERCI) in Piedmont, California. He has developed a software package, *DEFENDER*, used for budgeting, strategy documentation and selection, and reserve forecasting by more than 125 corporate environmental project managers on more than 2,500 sites. He founded ERCI in 1994, after serving as a guest scientist at Lawrence Berkeley Lab, product manager and business planner at FMC Corporation, and technical writer at OHM Remediation Services. Mr. Rosengard holds a B.S. from Georgetown University and an MBA from Northwestern University.

Judith Vangalio is a senior risk analyst at Vanasse Hangen Brustlin, Inc. and has fifteen years of experience in environmental consulting encompassing ecological and human health risk assessments, environmental assessments and impact statements, permitting, wetland delineation and restoration, habitat assessment and monitoring, and multimedia environmental sampling. She has worked under various regulatory agency guidance systems including RCRA, CERCLA, and state regulations. Ms. Vangalio received her B.S. degree in Ecology from the State University of New York College of Environmental Science and Forestry in 1985 and an M.S. in Toxicology from the State University of New York at Buffalo in 1987.

Cris Williams is a toxicologist and manager of risk analysis in Environmental Risk Management at Vanasse Hangen Brustlin, Inc. With over eleven years of environmental consulting experience, he specializes in conducting human health and ecological risk assessments for impaired properties, as well as providing litigation support in matters relating to toxicology and the health sciences. Dr. Williams has provided technical support for and management of all aspects of human health and ecological risk assessments for a variety of private clients (*e.g.*, chemical and automobile manufacturers, railroads, pharmaceutical companies, utilities, real estate developers, law firms) and public entities both nationally and internationally. Dr. Williams has published extensively in the fields of risk analysis, toxicology, and public health and has presented on these and other subjects in professional, legal, and public settings. He received his doctorate from the University of Wisconsin-Madison, and he is a member of the Society of Toxicology and the Society for Risk Analysis.

ACKNOWLEDGMENTS

The author and contributors wish to thank the following individuals for their help during preparation of this manuscript.

Randi Gonzalez, Gerry Jaffe, and *David Packer* of CRC Press/Lewis Publishers for their ongoing interest in and assistance with this project

Holly Forden of River Rock Design in Batavia, NY, for her assistance with many of the figures in the book

Jerold Bastedo and *Terri Courtemarche* of VHB for creating the original graphic for the cover of the book

Patty Chapman of VHB for editorial and formatting assistance

Kerry Weller of VHB for research

Jon Feinstein, MCP and *Robert Christman, PE* of VHB for allowing the time and space to complete the book

Finally, I must express my deepest appreciation to Joan, Nels, and Garth for their love, constant support, and patient understanding with the long hours required to prepare and edit a book while working full-time.

TABLE OF CONTENTS

DEDICATION

To those who taught me about science, the care of health and the environment within an urban-industrial landscape, and the realities of management:

J. Michael Daly, Ph.D.
Lloyd Darrow, Ph.D.
Jeffrey Davidson, Esq.
Karl H. Frantzen
Gerald Gallagher, Jr.
Raymond D. Harbison, Ph.D.
Lowell B. Johnson, Ph.D.
E.J. Kemnitz, Ph.D.
Mary Alice Kobovy
Lawrence H. Liebs
R. Gary Thurman, Ph.D.

Chapter 1

REALITY-BASED MANAGEMENT

Kurt A. Frantzen and Jerry Ackerman

I. INTRODUCTION

Why does a company or public organization become involved in managing environmentally impaired properties? Because of agents of change that, either individually or in combination, influence the entity's relationship to these properties. Some of these agents of change include:

- Direct Regulatory Action

- Releases of Hazardous Substances

- The Organization's Best Interest

- Third-Party Action Causing a Release

- Financial Considerations

- Community Activism

- Environmental Assessment Resulting from Transaction or Development Pressure

Once involved in managing environmentally impaired properties, many (if not most) organizations begin their activity with a common strategy, namely:

> *to do only what is required to minimize financial impact over as long a period of time as possible.*

This strategy rarely endures once agents of change voice concern (inside and outside of the organization). In so doing, they bring pressure to bear on the organization to address the environmental issue. This dynamic is especially true when there are health and ecological concerns regarding properties.

Cost-effective action is obviously desirable. However, a more productive course is possible with a strategy that maximizes the use of positive changes in regulations, potential site reuse and real estate options, and technological improvements to address or resolve adverse issues, achieve the organization's best interests, and ultimately improve the assets of the affected community.

We believe that the best way to achieve this goal is to implement the concept of *designing risk, wherein redevelopment or restoration is remediation.* When focused on maximizing the value of property and its influence on a community, it is easier to define risks to health and the environment and, in turn, design an effective response to a potential impact. The focus is to change a fragmented "losing" situation to one of mutual gain. Of course, the impacts of and consternation caused by internal and external political agendas may still have to be dealt with, requiring thoughtful diplomacy.

Sustainable development, now an ever-present paradigm, accentuates the concept of redevelopment as remediation. Such development can be defined as a coordinated action among business, government, communities, and individuals that leads to meeting present needs without compromising the ability of those in the future to meet their own needs (Robinson, 2000). As an integrating function, sustainable development is not and can never be the province of one individual or entity. While an organization *by itself* cannot deliver sustainable development, they can contribute to it through actions that:

- define essential needs,

- seize opportunities, and

- mitigate environmental threats.

Corporations are essential in this effort because of their financial and human resources, ability, and technology (Voorhees and Woellner, 1998). This is what we mean by *designing risk*: by defining, measuring, and adjusting (that is, *"remediating"*) the risk system (or impact of concern) it is possible to redesign that system or impact thereby ameliorating environmental liability, elevating the market value of a property, and restoring the property to improve the community's structure and function.

In the past, there has been too much focus on technical issues, with environmental remediation as an end unto itself and dealing with a concept of "removing every toxic molecule." Instead, the focus should be on integrated management that balances a multitude of issues, especially those that set the foundation for the remediation process contingent upon follow-on restoration, redevelopment, and reuse. To us, *decision-making*, which leads to the rehabilitation and productive reuse of impaired property, *is the essence of environmental protection*. The actual remediation of a site, while sometimes the most complex and resource-oriented process component, should not be the solitary goal. If other drivers are **not** planned for, addressed, and satisfied (especially stakeholder involvement issues), remediation becomes a very difficult management problem because it is severed from the "whole" context. This problem can scuttle any hope of remediation being cost-effective or progressive, and it can delay the redevelopment or reuse, as well as the accomplishment of fixing or correcting the impact or impairment.

A. THE NEED FOR AN ECOSYSTEMS PERSPECTIVE

Although we all know it intrinsically, nevertheless it is good to remind ourselves that urban ecosystems are complex and dynamic (Alberti, 1999 and Lubchenco, 1998). Preoccupied with economic prosperity throughout the Industrial Revolution, it took us a long time as a society to come to this understanding. Through the late 1940s and into the 1960s, we in the United States came to a large community consensus regarding the impacts of progress within the urban-industrial ecosystem with its implications of pollution. This consensus led to a political and policy decision embodied in the National Environmental Protection Act (NEPA) and similar laws at state and local levels. Quickly thereafter, our society came to a greater realization that not only did current progress and development pose a potential for environmental impact, but our historical activity did as well. On the heels of situations such as Love Canal, Valley of the Drums, and Times Beach came additional command-and-control laws intent on protecting human health by controlling the production, handling, and disposal of hazardous substances, hazardous materials, and hazardous wastes. At almost the same time, but somewhat behind in time series, came similar concerns about the threats and impacts to ecosystems and their inhabitants.

As with ecosystems, environmental issues are similarly complex and dynamic, especially when one is dealing with the interplay of chemical, biological, physical, and nuclear agents that may affect ecology and health. This complexity comes from an inextricable linkage between ecology and health (Lubchenco, 1998). The McElroy and Townsend (1996) working model of ecology and health informs the approach outlined in this book (Figure 1.1). Their model visually reminds us that the various parts of a system are dependent upon the other parts and that they are in continual interaction. The various spheres or units function as a singular whole (ecosystem) with relationships among the various populating organisms (including people) and their environments. This "interdependency" is important to remember because we too often attempt to mange things and solve urban problems based on immediate interpretations of cause-and-effect. Adverse health and ecological impacts never have single causes. Even so, one cause-and-effect is clear: *we affect the environment, and it affects us*. Moreover, the model further reminds us that adverse effects arise from physical, biological, or cultural imbalances, including changes in culturally constructed environments (such as buildings, streets, slums, and suburbs).

The McElroy and Townsend model provides a systems framework for considering action. Voorhees and Woellner (1998) argue that a complete systems approach is necessary to deal effectively with environmental issues. Thus, resolving environmental impairment issues involves more than just removing "toxic molecules." Instead, it involves individuals, populations, and diverse "systems"— biotic, abiotic, and cultural. This complexity requires that environmental risk management address a great breadth of issues to get things done. These issues include:

- defining the problem;
- determining what data and knowledge are crucial;
- eliciting input and support;
- clarifying the many forms and types of uncertainty and variability;
- determining how to inform decision-making processes;
- achieving a consensus;
- deciding on how to communicate findings and results; and
- working cooperatively to achieve a result that effectively addresses the environmental issue, but does so cost-effectively and in a manner that achieves safety and an improvement in the community.

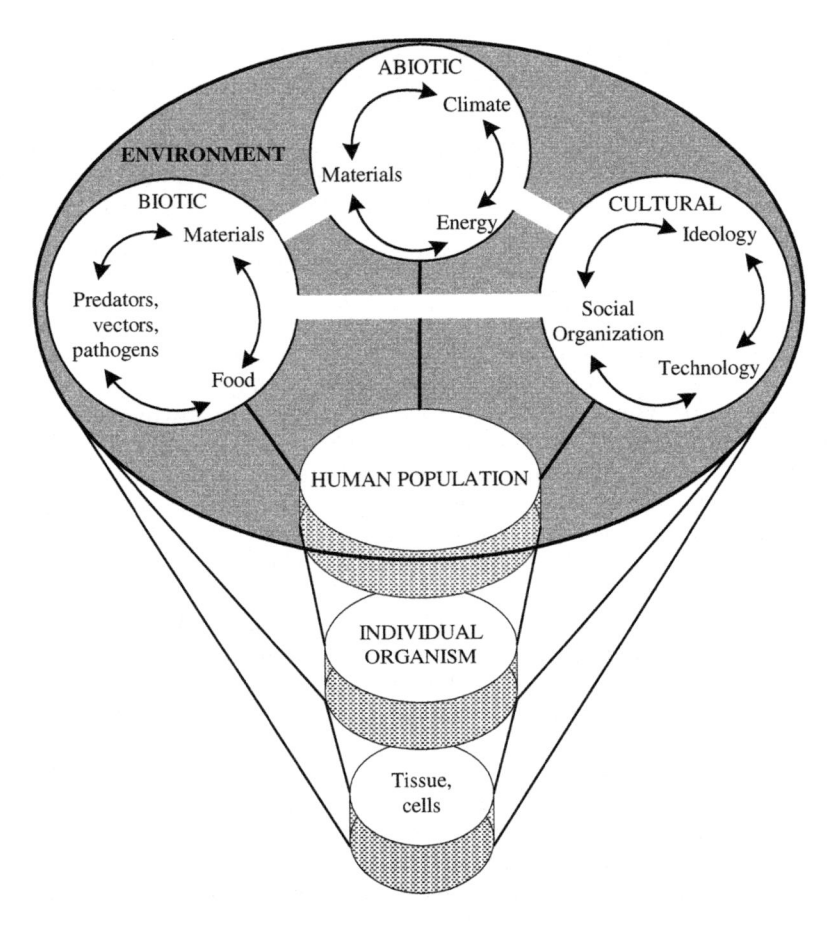

Figure 1.1 A working model of ecology and health (from McElroy and Townsend, 1996 with permission).

B. ENVIRONMENTAL RISK MANAGEMENT AND THE "SYSTEMS APPROACH"

Too often, the term "environmental risk" evokes thoughts of nothing more than a Recognized Environmental Condition or a source of risk, such as a particular substance. Others see it as a process of managing loss exposures (that is, financial losses covered by insurance, for example). As we will see, environmental risk and its management involve these and many other issues.

A.R. Wilson (1991) described "environmental risk" as a complex system, which includes:

- source(s),
- primary control mechanisms,
- transport mechanisms,

- secondary control mechanisms, and

- targets (people, sensitive habitats, or biota).

Incorporating McElroy and Townsend's model with Wilson's concepts results in a highly informative and useful schematic called a Conceptual Risk System Model (or CRSM) (Figure 1.2). Defining and understanding this system is crucial to effective environmental risk management because it helps property managers guide their strategy and tactics as well as guide science and engineering.

By applying the concept of *designing risk wherein redevelopment or restoration is remediation* within a systems approach, environmental risk management becomes an enlightened and productive way to *protect* human health and the environment. We emphasize the word "protect" because, when it comes to environmental impairment as "a peril," not only can there be loss exposure (to the organization), but *actual risk* to people. In this context, actual risk is a probability or chance of an occurrence (at some frequency and with some level of intensity) of adverse outcomes (sickness or death) to human health and ecological functioning.

Unfortunately, some stakeholders interpret "risk-based" management policies or processes touted by scientists, policy specialists, and industry as just a way to justify lower clean-up costs (by demonstrating that higher concentrations are not significantly "risky") and doing less than they (the stakeholders) believe is necessary. This perspective often leads community stakeholders to construe risk-based management as a smokescreen, a "cop-out," or a compromise of their personal health and the well-being of their community. At the same time, many in industry worry that the use of collaborative processes involving stakeholders will erode the primacy of science in risk management (Charnley, 2000). This perspective is often justified in that people often are convinced by the perception that the mere presence of chemicals or the possibility of exposure is tantamount to damage. Overcoming this polarization requires that organizations use a holistic systems approach that can build bridges to achieve environmental and economic objectives for the best interests of the community and its residents. Such an approach involves:

- Financial implications and one's ability to compensate for these implications through risk control (*e.g.*, pollution prevention, remediation, and institutional controls to mitigate or cut off exposure) and risk financing (*e.g.*, insurance).

- Concerns of the owner/seller or buyer (to limit regulatory and other legal liabilities).

- The requirements of regulators (seeking real protection of human health and the environment, not just a patchwork quilt of "risk-based" band-aids).

- The concerns and interests of the public (who desire and deserve a safe, sound living environment).

Loss exposures in business are dealt with via risk control and financing mechanisms. However, dealing constructively with environmental risk (and associated concerns of health and ecological impacts) from a strategic management perspective requires use of "the hard currency of knowledge." As others, most

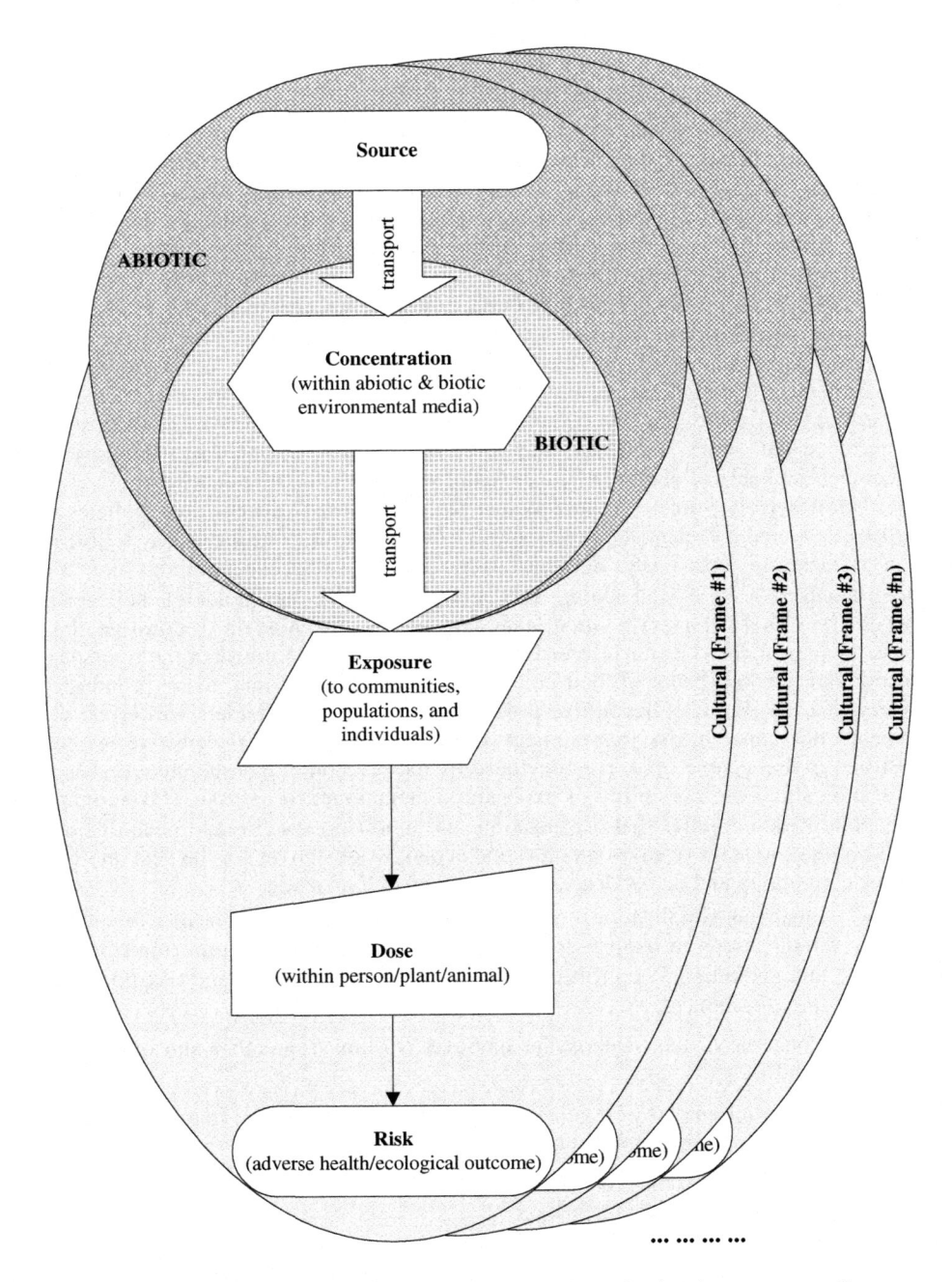

Figure 1.2 The Risk System resides within abiotic, biotic, and cultural systems, as well as differing personal (and institutional) frames of reference within those cultural systems, producing multiple cultural frames.

notably Stephen Covey (1989), have said, we must "...begin with the end in mind!" In other words, if you do not know where you are going, you (in effect) are not going anywhere, or if you do not know where you are going, any road will get you there.

What differentiates the environmental risk management approach described in this book from other approaches or books on the subject is:

- First, a belief that the right knowledge, not just more information, leads to less ignorance and a greater appreciation of the business risks one is facing. Corporate managers receive considerable advice of all kinds (financial, legal, scientific, insurance, political, economics, and moral). Much has improved over the past 20 years in terms of knowledge and understanding of these subjects. Nevertheless, the matter of making decisions about environmental impairment, based on risk to health or the environment, remains challenging. Perhaps the situation has improved somewhat with the advent of Risk-Based Corrective Action (RBCA) and recent voluntary regulatory approaches. However, corporate managers still need better tools to:

  - improve the decision process,

  - better manage the use of science (especially the art and science of risk assessment), and

  - improve communications and collaboration with the public.

- Secondly, a focal endpoint on economic redevelopment appears. All other issues pertaining to a site—whether the surrounding infrastructure, environmental conditions, demographic profiles, etc.—are weighed in terms of how they affect the economic status or candidacy of an impaired property. Obviously, some pieces of real estate are never a bargain, regardless of their sales price or environmental condition. The strength of their status or candidacy centers on returns-on-investment associated with site use/reuse. While this logic may be readily apparent, values placed on properties frequently do not account for variables that have real limitations associated with remedial and restoration costs. Conversely, basic site characteristics (like location, adjacent properties, existing infrastructure, etc.) may provide a value sufficient for a potential buyer to invest time and money in initial transaction and due diligence expenses. There are, of course, some sites in which their environmental restoration may be viewed as a catalyst to social benefits (community health and welfare or safety, etc.) more than economic returns. However, even in these cases, economic returns, although not immediately apparent, can be enormous—not for the site unto itself perhaps, but in relationship to its economic and environmental impact on surrounding properties, the local vicinity, and the community. For example:

  - **Sustainable Growth**—Environmentally impaired sites that appear to be of no or less-than-no value can create safe, clean, and easy access to major, economically viable sites—assets like large waterfront parcels which in turn attract high-end, private-sector investment. As those large, economically viable sites are reused, they induce further economic viability (*i.e.*, multiple, additional return) into other nearby sites once called financially "marginal" or "upside-down."

- o **Neighborhood Value**—In other situations, transforming financially marginal or upside-down sites into new enterprises or even green spaces increases the value of residential communities and nearby commercial enterprises. This interdependent structure for economic well-being energizes further entrepreneurial and economically upgrading activities.

- o **Economic Synergy**—Ironically, some helpless sites, those with an apparent negative value, can nourish other sites with additional value to create a holistic worth and economic transformation for an entire city region. As research information is compiled, evidence is mounting to demonstrate that marginal or upside-down sites can create an exponential value that significant, economically viable sites cannot always do all on their own (Ackerman and Soler, 2000).

II. RISK-BASED ANALYSIS

A. INTRODUCTION

Before defining what we mean by Risk-Based Analysis, it is important to define several associated terms and make certain distinctions in order to bring those familiar and unfamiliar with the science and art of risk analysis onto common ground.

1. Risk Analysis

Risk analysis is a two-step process:

1. Evaluating (qualifying and quantifying) risk.

2. Making (policy or reuse) decisions based on the evaluation together with other input (*i.e.*, the "whole picture" as previously mentioned).

The process involves communication among those involved in the process, including, as a major focal point, those potentially exposed to risk. While using science, risk analysis is not wholly objective, empirical, or fact-based. It is simply an analytical tool for evaluating the magnitude and severity of risk. It uses information that cannot be known with certainty and only produces an estimate, never an exact prediction. This limitation is particularly true with environmental issues because data are often sparse and scientific theories explaining hazards, exposures, and effects are tentative, if they exist at all. Risk analysis applied to the environment requires choosing among plausible assumptions and competing theories, as well as regulatory policy decisions, to bridge gaps. For these reasons, as well as the subject complexity and unfamiliarity to most people, risk analysis is subject to challenge.

2. Risk Assessment

Risk assessment is the process of estimating the likelihood that a given effect will result from a specific presence, action, or activity (where likelihood is a probability and interpreted as the portion or fraction of time a consequence might be observed). Concerning toxic substances, risk assessment involves determining the likelihood of release (exposure) and the resulting consequence (hazard). The National Research Council/National Academy of Sciences has explained the assessment of risk in four steps (NRC, 1983) as listed herein:

- Hazard Identification
- Toxicity Assessment
- Exposure Assessment
- Risk Characterization

Chapter 3 and Appendix A discuss these steps in detail.

3. Risk Management

Risk management is the process of identifying, evaluating, selecting, and implementing actions to reduce and control risk to human health and the environment. The process involves considering scientifically sound, cost-effective actions to reduce or prevent risk, while taking into account social, cultural, ethical, political, and legal considerations (PCCRARM, 1997a, p.1).

4. Risk Communications

Risk communications is the exchange of information about health and environmental risks among risk assessors, risk managers, the public, media, interested groups, and others. Chapter 3 and Appendix C discuss this crucial process in detail.

5. Risk-Based Corrective Action

Risk-based guidance and decision-making regarding petroleum-contaminated sites and others contaminated by hazardous substances is steadily gaining acceptance within the overall environmental industry, including the United States Environmental Protection Agency (USEPA) and state regulatory agencies. RBCA is an environmental management tool developed to streamline typical regulatory models and more effectively guide the assessment, remediation, and closure of (originally) petroleum contaminated and now even hazardous substance and hazardous waste site(s).

RBCA refers to the American Society for Testing and Materials (ASTM) Subcommittee on Storage Tanks standard [E-1739, ASTM, 2000] *Guide for Risk-Based Corrective Action Applied at Petroleum Release Sites*. It also refers to the *Standard Provisional Guide for Risk-Based Corrective Action* (PS 104-98, ASTM, 1998). These ASTM standards are an example of risk-based decision-making methods incorporated into corrective action programs consistent with USEPA and state agency policies and regulations. The goals of RBCA are:

1. protectiveness of people and resources,
2. practical and cost-effective application using limited available resources, and
3. consistent and technically defensible processes.

RBCA is an iterative procedure providing a basis for the following:

- Identifying the initial source release and the necessity for and duration of emergency response initiatives, oftentimes referred to as interim remedial measures (IRMs);

- Focusing the collection of high-quality and reproducible environmental site assessment data that will identify all potential exposure pathways, receptors, and source mechanisms, and adding credibility to the use of alternative real-time field data collection methodologies;

- Categorizing or classifying a site, or portfolio of sites, according to the perceived threat/risk presented to human health and the environment;

- Assisting in the calculation and establishment of site-specific objectives and targeted cleanup levels;

- Determining what, if any, further action (corrective/remedial, continued compliance monitoring, etc.) is required to bring a site(s) to the point of no further action; and

- Deciding on the level of oversight provided to cleanups conducted by responsible parties.

B. OVERVIEW OF RISK-BASED ANALYSIS

Environmental risk management requires consideration of not just hard, objective, scientific fact but thoughtful consideration of their subjective nature amid various personal and cultural perspectives (Figure 1.2). This situation and requirement exists because, as stated by Tillich (1968),

> *"Reality precedes thought; it is equally true, however, that thought shapes reality."*

As you read this postulate, you may be thinking that when the environmental impairment is on *my* property, then the issue is only a simple evaluation of "loss exposure" and meeting regulatory requirements. This impairment should not concern anyone else (other than regulators, financial analysts, lenders, the Security and Exchange Commission, and stockholders). However, "loss exposure" takes on an entirely different perspective when others (stakeholders) think about situations involving *their* exposure (to *your* or *my* property's "impairment") and *their* loss (of welfare, health, life, or just impairment of *their* "quality of life"). As Voorhees and Woellner (1998) point out, those engaged in environmental risk management need to understand that:

- The unfamiliar is less acceptable than the familiar.

- The involuntary is less acceptable than the voluntary.

- The undetectable is less acceptable than the detectable.

- Perception of unfairness is less acceptable than fairness.

- The dramatic and memorable impacts of the adverse are just unacceptable.

- Stakeholders want reductions in risk – not risk estimation, but foremost they want their fears and concerns validated.

Let us combine the quote from Tillich with these points. People trained in science and business use objective and subjective reasoning to shape reality, that is, they grasp reality through understanding and express it with their mental tools applied to formal disciplines. In reacting to this "world," they shape it by transforming it into a gestalt, a living structure. Similarly, community stakeholders do the same thing but with mental tools applied to informal disciplines; and this process results in different gestalts. Industry hopes the primacy of science will prevail in environmental risk management, but how can it when the realities of those involved—the gestalts—are disparate?

Therefore, managers must deal constructively with the objective and subjective multi-dimensional reality of "risk systems" (Schrader-Frechette, 1991 and Kervern, 1995). To do so requires a management process that uses the currency of knowledge, not just finance or insurance, to apply a fair process following the principles of engagement, explanation, and expectation clarity (Chan Kim and Mauborgne, 1997). This policy or principle is part of incorporating stakeholder values into corporate environmental decisions (Earl and Clift, 1999).

Risk-Based Analysis provides systematic guidance for developing necessary information and insight to the objective and subjective aspects of an environmental impairment and potential impacts to human health or the environment. It helps the manager understand how the impairment becomes a loss exposure (via property, personal, and liability loss, which combine into net income loss). Thus, Risk-Based Analysis *is not* "risk assessment" in the formal definition of the term, although such assessment is a part of the analytical process. Rather, Risk-Based Analysis is a tool to help **you**, the manager (whether you are a Risk and/or an Environmental Manager), better implement, direct, and use risk assessment and risk communication in concert with engineering and the usual organizational resources to achieve a managed risk solution. The purpose is to empower you to better influence the multi-component and multi-party decision-making processes and negotiations involved with (environmental) risk management. We believe that Risk-Based Analysis is a practical approach that can improve the interface between you, the manager, and others involved in performing the sub-disciplines of risk analysis (*i.e.*, assessment, management, and communication) and other technical fields within and beyond your organization. It helps the manager develop knowledge and understanding to guide internal decision-making, as well as contribute to external decision-making processes. In so doing, it should help you make the processes more responsive to your own business calculus.

Why is this important? The management of environmentally impaired property involves grappling with many different concerns, most notably:

- societal concern over adverse health effects and ecological impacts,

- business concern over the associated liabilities,

- business concern over their image as concern over environmental contamination within society has risen, and

- business concern over the financial implications.

We perceive that the definition of risk drives all of these concerns. In other words, the potential of an environmental impairment to cause harm informs the qualification and quantification of the risk system. Additionally, the significance and importance of the situation (the impairment and its perception) contextualize this definition. Finally, the uncertainties associated with the data, information, and models have great importance in defining and making decisions about risk.

As briefly described in Table 1.1, Risk-Based Analysis is an integrative technique with five progressive, knowledge-building value points to guide the management, collection, and analysis of risk information. The procedure provides a system to help you minimize cost and potential liability by articulating cleanup within the context of a property's socioeconomic value, planned redevelopment approach, and intended reuse(s). It provides the knowledge foundation critical to making strategic plans for managing a site and its impairment.

In contrast to RBCA, Risk-Based Analysis is not a model for a regulatory approach. Rather, it is a five-step tool to help inform your management process throughout its operation for a particular impairment or property. In so doing, Risk-Based Analysis provides:

- Insight into the objective and subjective issues related to loss exposure identification and evaluation (from the planning stages through site characterization and evaluating the nature and extent of contamination);

- Aid in deciding whether corrective action is necessary;

- Support in evaluating and selecting risk management alternatives;

- Information for the implementation process involving development of cleanup thresholds, evaluation of the safety of remedial implementation (the risk of remedy issue), and suggested verification techniques; and

- Guidance to the post-remediation monitoring process, including assurance of the effectiveness of institutional controls.

More importantly, how much data collection or inquiry is necessary to evaluate an environmental condition? It depends upon the issues, decisions, and level of remaining uncertainty acceptable to you and others. Unfortunately, consultants' conclusions are (at best) opinions, usually only advice, and occasionally wrong. Environmental site assessments (regardless of their phase), risk assessment, feasibility studies, etc., are exercises in judgment, based on science, engineering, and experience. Risk-Based Analysis gives you a method for designing, testing, and appraising (managing) their work in the light of your business needs and financial calculus.

C. RISKS AND REWARDS

Environmentally impaired properties and especially Brownfields sites are a risk—financially and, to one extent or another, environmentally. The *Brownfields Revitalization Process: Environmental & Economic Fusion* approach of Ackerman, *et al.* (1998) shown in Figure 1.3 has a different focal point than most approaches to these issues, that is, **economic redevelopment**. We all want to see our communities grow and prosper, and our urban ecosystems flourish as healthy places to live. The problem is how to cope in a (financially) manageable way with that risk, to meet

Table 1.1

Risk-Based Analysis:
Five Progressive, Knowledge-Building Value Points

Problem Formulation

First, define the problem; specify needed resources, deadlines, and scope. Use conceptual models to guide definition of source (cause), effect, and the many influencing factors. Establish the boundaries and operational context of the problem and the associated impairment or risk issue(s). Develop a preliminary model of the decision-making process and identify data needs to inform that process and define the necessary quality of data (*i.e.*, if you collect or calculate it, will it convince?)

Situational Analysis

Identify, understand, and integrate the needs and objectives of others within the regulatory, political, and socioeconomic aspects of the property and their roles in risk management decision-making.

Risk Assessment

Quantify and qualify the nature, frequency, and intensity of risk. Set the scientific data and findings in redevelopment/reuse contexts.

Risk Management Option Development

Depending on the problem and its situational context, address what options are available to scientifically and justifiably explain away the reputed risk or impairment, cut-off exposure pathways (and therefore risk), or permanently reconstruct the "risk system" (*i.e.*, source or effect) so that it no longer exists, or is quantitatively reduced in magnitude by a significant and sufficient degree. In addition, to help influence outcome options, develop your risk mitigation scenarios within the context of redevelopment and economic revitalization. It may even be worthwhile to develop a short- and long-term amortization of risk over a sufficiently long planning horizon to better contain costs and land use.

Risk "Argument"

In this step, develop a convincing communications approach to achieve optimal, "mutual gain" solutions by integrating property value, environmental risk or impairment, the situational context, risk-management options, and decision-making frameworks. The risk information and preferred risk management option are formulated within a communications program by which it is presented and, ultimately, negotiated into an approach acceptable to all.

those needs. The best place to start the process is at the beginning: by visualizing the endpoint, and this is where Risk-Based Analysis comes into play.

Risk-Based Analysis helps you place the environmental impairment (*i.e.*, the contamination constraint) in perspective and, in so doing, points the way to a cost-effective management solution. It informs what in essence is a restoration process and thereby strategically transfers environmental engineering efforts into the redevelopment/reuse planning efforts (recall the *designing risk* concept). This transposition leads to the effective application of remedial tools into planning, site/civil engineering, transportation, and other activities to achieve exposure (and risk) mitigation. In this context, *redevelopment and restoration is remediation.*

The application of Risk-Based Analysis is not the justification of less cleanup. Rather, its purpose is to build an environmentally and economically sound approach that people (site neighbors and other stakeholders) will agree as being safe and justifiable. Therefore, the purpose of Risk-Based Analysis is achieving holistic

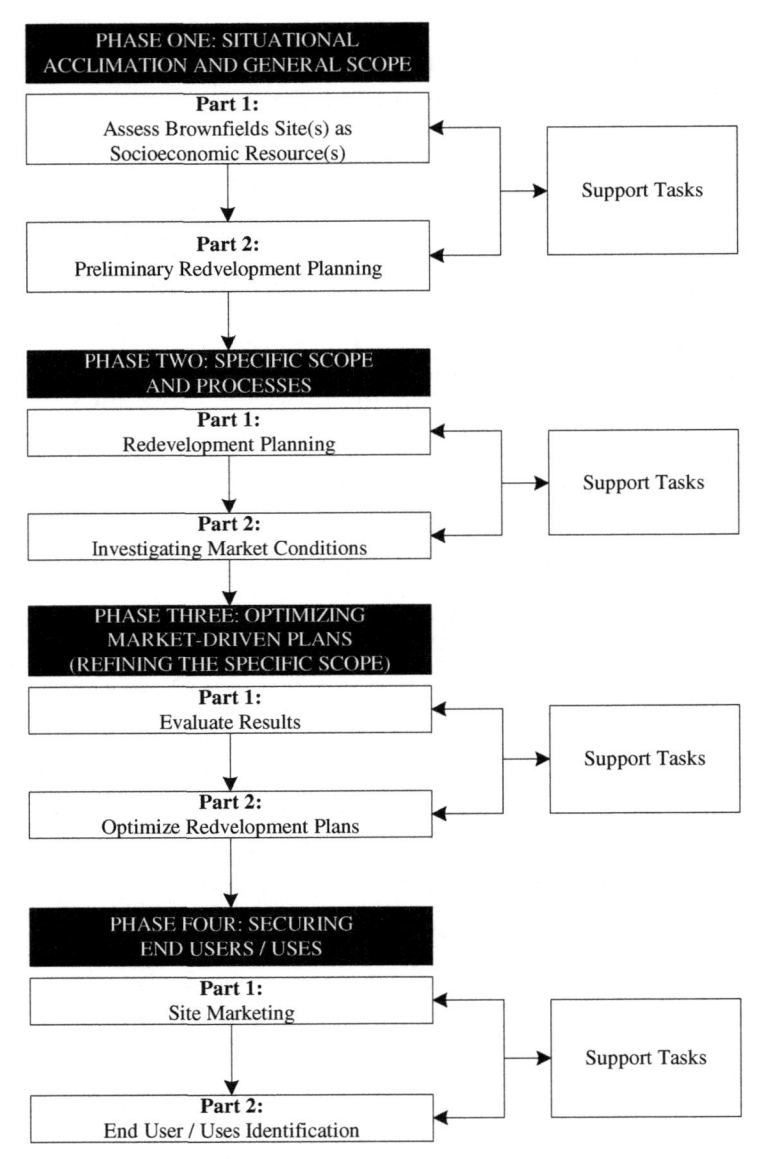

Figure 1.3 The Brownfields Revitalization Process: Economic and Environmental Fusion (after Ackerman *et al.* 1998, used with permission).

benefits (in the context of the dispute resolution approach articulated by Susskind and Field, 1996) for:

- The party responsible for the environmental impairment (*i.e.*, less expense and lower liability),

- The regulatory agencies (*i.e.*, another property cleaned up), and, most importantly,

- The community (*i.e.*, safety and improved socioeconomic circumstances).

What makes Brownfields deals and the reclaiming of other environmentally impaired assets insurable, worthy of financing, and ultimately achievable? The development of a practical and cost-effective risk management (read "cleanup") plan, obtaining community support, navigating the complex regulatory approval processes in a timely manner, and executing that management plan.

The information, knowledge, and understanding that result from Risk-Based Analysis supports the evaluation of redevelopment options, informs the "do or don't do" or "buy or no-buy" decisions, and helps you answer the question "why" with respect to decisions. This approach, when used early in your environmental, due diligence, and/or management processes, leads to concrete intelligence concerning the questions:

- If we want to buy the property, are we looking at good real estate, and what can we do about it given its environmental constraints?

- If we already own it, what can we do about this issue?

Perhaps these questions are an oversimplification of the environmental and economic fusion process, but they underscore evidential objectivity. Essentially, managers must act knowing what financial, political, and environmental margins they are dealing with when deliberating on committing valuable resources to a potential Brownfields redevelopment project, or resolving an impaired property asset. Risk-Based Analysis is a credible tool for clarifying and dealing with the uncertainty associated with the environmental risks of a property in order to reap the rewards.

III. OUTLINE OF CHAPTERS

This book advances the Risk-Based Analysis concept described in this Introduction. Through the course of three chapters, the context of the approach is set out and explained. Additional background and supplemental information is included in several appendices.

A. ENVIRONMENTAL RISK MANAGEMENT

In the second chapter, we look at the issue of making decisions about risks to human health and the environment. As business risks, environmental issues are unique and more complex than classic risk management loss exposures, and it is the decision-making about the environment that is the essence of its protection.

We also attempt to integrate environmental sciences, specific technical disciplines, corporate finance, legal affairs, and corporate communications and image. These different areas see the risk associated with environmentally impaired property from their own paradigms, just as individuals perceive risk differently. If there is to be effective risk management, we need to understand how the various definitions of risk come about: who defines the risk, and when. The chapter concludes with a discussion about the matter of risk and rationality and returns to the matter mentioned earlier, that is, a need for a systems approach.

B. INTERFACING THE ASSESSMENT, MANAGEMENT, AND COMMUNICATION OF RISK

The decision to manage environmental risk unleashes a process involving a knowledge-based network of individuals and entities, all with their own perspectives, agendas, and issues. Sound management principles first require the assessment of environmental risk (both its quality and quantity). This interesting job comes first, and, therefore, everyone hates it because everybody thinks their perspective, agenda, and issues are the most insightful, accurate, and preeminent. Here is where the problem solving starts, that is, through information development.

In contrast, risk management is the job that everybody wants (or at least thinks are capable of doing). Here is where the problem solving gets interesting! This interest comes about because risk management is where real alternatives are explored, weighed, and decisions made. However, who is the real risk manager, and should you inform (that is, influence) the manager and if so how?

Finally, there is risk communications. Here is one job that nobody wants! This "avoidance" tendency is because, the old saying goes, "when all is said and done, more is said than done." Despite all we may know, and will review in this book, we still prefer to apply the classic form of *Decide-Announce-Defend* (or the DAD approach) to most environmental issues. There is change coming, which we embrace and encourage herein, that involves a more inclusive, fair process to ensure procedural justice. This procedure is called the *Define-Agree-Implement* approach. The RBCA approach is good, but it lacks something for managers and communities living life in a knowledge economy: it does not interface well (although there are some signs of improvement).

C. THE PRACTICE OF RISK-BASED ANALYSIS

In the fourth and final chapter, we detail the Risk-Based Analysis approach and its four steps as briefly described in Table 1.1.

D. APPENDICES
1. Appendix A. Evolution of the Risk Paradigm

Appendix A provides an overview of the major evolutionary steps in the policy and technical aspects of the risk paradigm primarily across the United States—as expressed through documents of the National Academy of Sciences and the USEPA, —but also in certain select states, as well as in some other countries. The discussion looks at how the paradigm pertains to environmentally impaired property and how it has led to defining risk, justifying the basis for remedial action, and providing a technical definition of what is clean.

2. Appendix B. Evaluating Financial Liability Implications of Environmental Risks

Better financial liability modeling and applying financial analysis to remedy selection decisions are essential because:

- Of the costs involved in remedying environmental risks, the impacts that occur to current operations, and the potential for shocks from large reserve charges to shareholders.

- Buyers and sellers value property assets differently (whether they are "clean" or not) and mergers-and-acquisition activity affects valuations of not only assets but liabilities as well.

- Environmental consultants and remedial contractors know of the variations among environmental managers and that companies pay widely divergent amounts to achieve the same level of protection of health and the environment.

This appendix discusses some of the things that you can do to assess better the financial implications of how risk is defined and how to better affect your environmental risk management program.

3. Appendix C. Risk Communication Basics

While much about dealing with environmentally impaired property involves science, engineering, technology, and sound business practice (legal, finance, etc.), communicating about it is art. This appendix provides the rudiments to risk communication.

- Three rules of managing the public issues involved with environmentally impaired property.

- Four priorities that must be addressed when implementing risk communications pertaining to environmentally impaired properties.

- A series of decision priorities, which assure that each act of communication is based on an efficient consideration of all issues that impact what a company says and how it says it.

- A developed communications strategy and plan coincident with the development of a management strategy for the impaired property itself.

4. Appendix D. Risk-Based Analysis Workbook

The Risk-Based Analysis (or RBA) Workbook is a tool to help managers plan and implement an environmental risk management project following the protocol described in this book. There is a twofold objective for the workbook:

1. As a manual, that summarizes the RBA protocol.

2. As a record of work related to an environmental risk management project.

The workbook provides a checklist and guidance for performing each of the five steps:

- Problem Formulation

- Situation Analysis

- Risk Assessment

- Risk Management Options Development

- Risk Argument Development

5.   Appendix E. Acronyms and Glossary

This appendix provides a listing of the many acronyms used throughout the book, as well as a glossary of terms.

Chapter 2

ENVIRONMENTAL RISK MANAGEMENT

Kurt A. Frantzen

I.   INTRODUCTION

"By day, we work with statistics; in the evening, we consult astrologers and frighten ourselves with thrillers about vampires. The abyss between the rational and the spiritual, the external and the internal, the objective and the subjective, the technical and the moral, the universal and the unique, ...grow deeper." (Havel, 1996)

The quote above comes from an article based on a speech given by Vaclev Havel titled, "The Need for Transcendence in the Postmodern World." In the past decade at least, realization has grown for the need to transcend boundaries in order to end the insularity of our thinking and ourselves. Society in general seems in search of a strategy to integrate knowledge. New cooperative and cross-disciplinary approaches are emerging in many areas. Likewise, integrative approaches are essential in environmental problem solving.

Interestingly, the management of environmental risks often occurs within an urban ecosystem where hazardous substances and hazardous waste exist in a milieu of daily chemical exposures for us all (Ames and Gold, 2000 and Hoddinott and Lee, 2000). As Ames and Gold opine, "[e]ven Rachel Carson was made of chemicals," their point being that people who are concerned about environmental contamination need to keep things in perspective in that humans receive exposure to many chemicals in the course of their normal lives. Nevertheless, this pivotal issue involves more than just identifying and evaluating a loss exposure associated with an impaired piece of property. Environmental risk management involves the definition of potential loss in terms of hazard, harm, risk, and cost. It is a multi-dimensional effort using input from many sources (within and without the corporation), disciplines, and perspectives. Thus, its real problems are trans-boundary, requiring a holistic approach to yield solutions that integrate corporate needs with those of society.

In this chapter, we explore the integrated nature of environmental risk management. It encompasses environmental sciences, specific technical disciplines, corporate finance, law, and corporate communications. Each different area of expertise sees the issue of risk from its disciplinary paradigm and gestalt—just as individuals perceive risk differently. Therefore, if a corporate manager is to manage effectively a company's environmental risks, then it is necessary to understand how various definitions of risk arise. Just as important is the ability to identify those who are critical to the definition of their company's environmental risks, and when that definitional moment occurs within and without the company.

II.   HOW RISK IS DEFINED

Although the world is chemical in its totality, human activities occasionally lead to increased, even excess, concentrations of some chemicals in the environment. Modern life has increased the extraction of naturally occurring chemicals, their manipulation, and use.   Over the years humanity has increasingly applied the chemical arts to synthesize a vast chemical armamentarium serving many diverse human needs and interests. These activities have led to:

- societal concern over adverse health effects and ecological impacts;

- business concern over the liabilities associated with the use of the chemical arts, the release of chemicals from commercial activities into the environment, and the subsequent implication to stockholder value from those liabilities; and

- business concern over corporate image and perceptions to stockholders, clients, and the public as concern within society has risen about environmental contamination.

Risk management is defined by Rejda (1992) as:

"Executive decisions concerning the management of pure risks [circumstances of only loss or no loss], made through systematic identification and analysis of loss exposures and search for the best methods of handling them."

Rejda speaks of the issue of identification and analysis of loss exposures. However, loss exposures from environmental impairment are more complex than other types of liabilities.   One of the primary complications is the definition of the risk to be managed. In this section, we will look at efforts to define the risk or impairment arising from environmental contamination by:

- corporations,

- regulators,

- those who strongly believe in the effectiveness of science-/risk-based decision-making, and

- those who are more cautious.

A. ENVIRONMENTAL CONTAMINATION: A CORPORATE CAUSE CÉLÈBRE

The corporate manager must deal with a diverse set of opinions and approaches about environmental risks.   The discussion here identifies the various groups involved in defining risk for the company: legal, corporate relations, and finance. The starting point of it all is identifying and defining the impairment (the "hazard"), the environmental risk system (refer to Chapter 1 Section IB and Figure 1.2), and the actual peril or cause of loss, which is human and ecological risk.

1. As a Health and Ecological Peril and Concern

Observation of acute (immediate or short-term) health effects was common before the nineteenth century. Beginning in the twentieth century and concurrent with improved public sanitation and health and longer life, we began to gain more experience with chronic health problems due to many different causes. Regardless of the cause-and-effect relationship, the increased use of, and dependency on, chemicals leads to an increase in the potential for exposure. This increased potential thereby leads to:

- increased risk (or *objective probability* using insurance terms, p. 6 of Rejda, 1992) of adverse health effects and environmental impacts and, perhaps in some cases, the actual prevalence of adverse health effects and environmental impacts; and

- fear of adverse health effects and environmental impacts (or *subjective probability* using insurance terms, p. 6 of Rejda, 1992).

Those closely intimate with a concern about health or environmental impact may very well judge that the matter is a high priority because:

- People will judge the impact's significance in terms of proximity of time and space to themselves, their living space, their community, and in terms of the range of people, things, and biota/habitat actually or potentially affected (Llewellyn, 1998).

- They judge its importance based on issues of relevance to their understanding of public health, ecological soundness and functionality, ethno-/social-cultural well-being, and socio-economic vitality. Thus, the benefits of correcting the situation, especially to them, justify the cost because they see themselves as the ones at risk (Llewellyn, 1998).

- Regardless of the uncertainty of the data and its variability or the uncertainty of the actual/supposed consequence(s), the peril or concern is an objective reality to them (*i.e.*, thought shapes reality).

These aspects serve to define the environmental impairment (due to chemical contamination) as "a peril," that is, a cause or source of loss (p. 7 of Rejda, 1992). The impairment is not just an insurable loss exposure to the company, but *actual and perceived risk* (or perhaps more accurately stated as a chance of loss [Rejda, 1992 p. 6]) to somebody or something as well.

It is important to point out here that the use of the term *risk* in classical risk management and in the insurance industry is different from that used by scientists, regulators, and others involved in the environmental industry.

- From a classical risk management/insurance perspective, *risk* is the "...uncertainty concerning the occurrence of a loss" (Rejda, 1992 p. 5).

- In contrast, a common definition of *risk* from an environmental industry and regulatory perspective is the probability that damage to life, health, property, and/or the environment will occur because of a given hazard.

The insurance industry focuses on uncertainty. From their perspective, the emphasis on *probability* actually reflects the chance of the loss itself, not the level of impact or actual damage that may arise, as mentioned above. Reading these definitions also reveals differences in the definition of the term *hazard*. From the environmental industry and regulatory perspective, *hazard* is the potential to cause illness or injury. From the insurance perspective, those things or conditions creating, or increasing the chance of a loss, are *hazards* (*op cit.*). For the purposes of this book, while we will have an emphasis on probability, we will use the term *risk* from the common environmental regulatory/industry perspective. Therefore, in this context and for the purposes of this book,

- *Actual risk* is the probability or chance of the occurrence of adverse outcomes (sickness or death) to human health and ecological functioning at some frequency and with some level of intensity.

- *Perceived risk* is risk as defined by its importance and significance to individuals, the community, and society.

These two types of risk define *peril*—the cause of loss (Figure 2.1). This peril drives all other business concerns or risks (legal liability, financial liability, property value diminution or damages, and negative stigma and perception of the corporate image) that arise from an environmental impairment. If the impairment or peril did not exist or was defined differently then the responsible party, the regulatory agencies, and other stakeholders could, and most likely would, react differently.

2. As a Legal Concern

As previously discussed, many environmental and occupational laws have been enacted over the last thirty or so years in the United States to protect the public and workers from adverse health effects, as well as impacts to the environment, from chemical exposures. These laws require mitigation and cleanup of the release or spillage of oil, hazardous substances, or hazardous waste. The legal liabilities associated with such activity include involvement by federal and/or state agencies either leading the cleanup or operating in an oversight capacity, natural resource damages, and third-party claims, including "toxic tort." Additional liabilities are associated with litigation against other potentially contributing responsible parties and insurance carriers. The liabilities stem from the peril or actual and perceived human and/or ecological risk described above and how they are defined (Figure 2.2).

3. As an Image Concern

The advent of an environmental impairment or peril poses another liability to the potentially responsible party, that is, to its image. As discussed in Appendix C, the responsible party is dealing with powerful and unprogrammable emotions when informing people that a contaminated site exists in their community and that it must be remediated to protect their health and environment. It also is dealing with a complex of interest group interactions. The stigma associated with an environmental impairment (such as chemical "contamination") and the variety of people's perceptions of it, as well as people's perceptions of the responsible party itself, are dynamic. The responsible party must discover and comprehend the varied interests, wants, and needs of the various parties involved and then balance its relationship to the community, the regulators, and the many political interests of that locale.

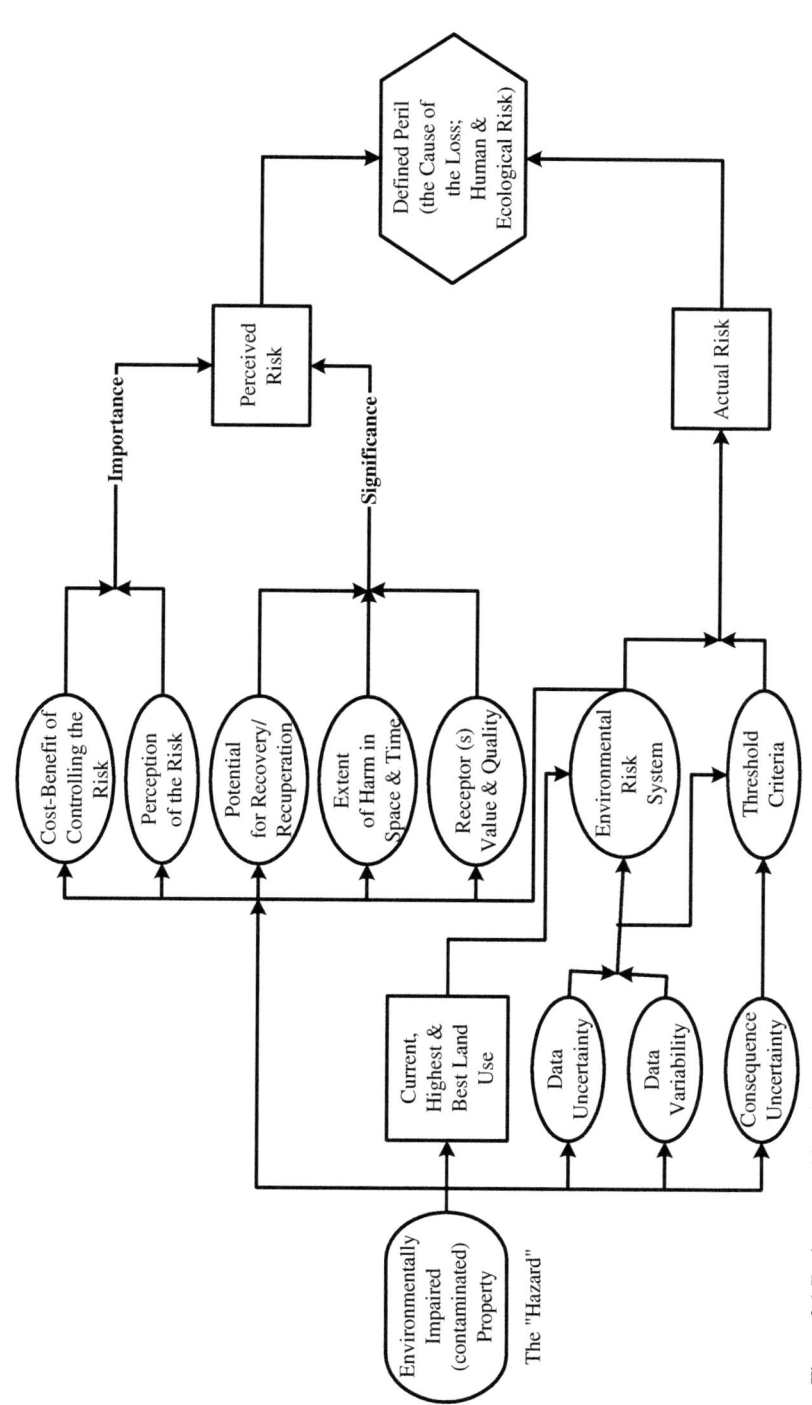

Figure 2.1 Environmental impairment due to chemical contamination results in the definition of an Environmental Risk System (defined in Chapter 1, Sections A and B and Figures 1.1 and 1.2). This yields actual risk and evokes perceived risk through subjective means as defined by the importance and significance of the risk, which together define the health and ecological peril or cause of loss that drive all other business concerns.

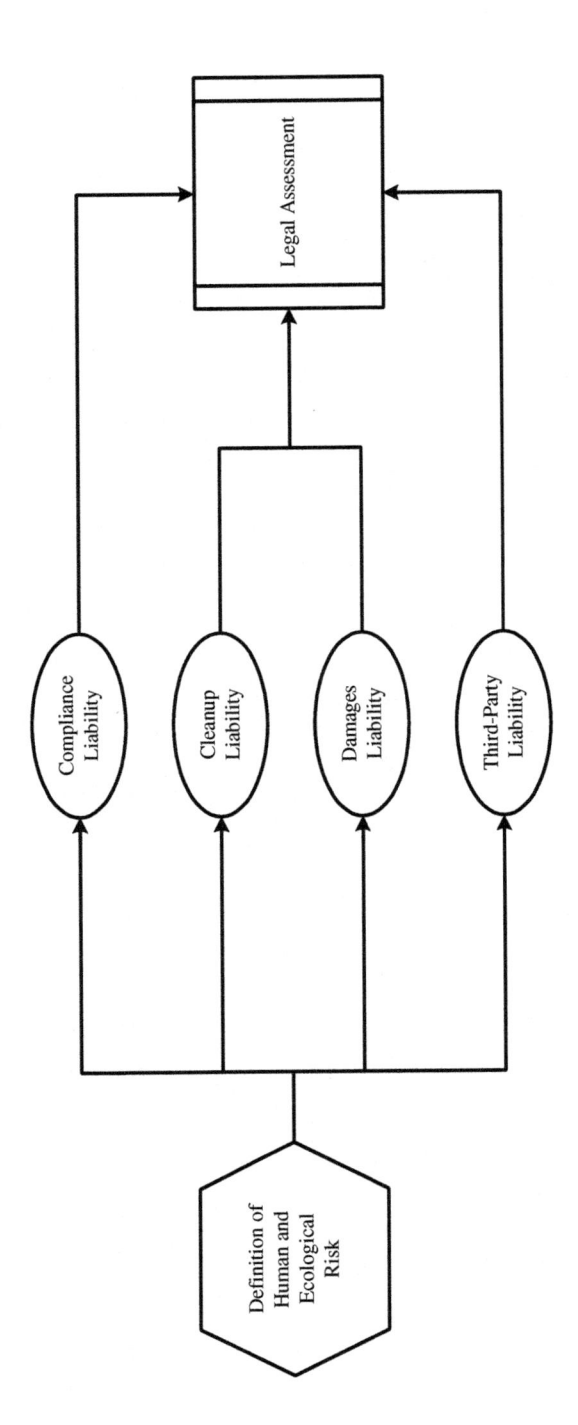

Figure 2.2 The definition of human and ecological risk is a concern, a peril, and a cause of loss for the company and thus has various legal implications requiring assessment. This figure summarizes some of the crucial types of liability requiring assessment, all of which are driven by the definition of risk.

Liability to the company's image arises from the definition of the peril as an actual human and/or ecological risk, which is simultaneously colored by stigma, as well as individual and group perceptions (Figure 2.3).

4. As a Financial Concern

Environmental risks are one of the major contingent liabilities facing companies today. This type of contingent liability causes several types of financial concern.

- The management problems associated in dealing with the issue.

- A detrimental condition to the property in question causing damage to the property's financial value.

- The question of whether the environmental risks produce sufficient property impairment limiting use of the site and/or increasing other business costs.

- The accounting profession has similar tests to check if a company has adequate reserves for each liability type, but there is no standard method for calculating the reserves of any contingent liabilities. So there is the question, which is of paramount importance to stockholders, of whether the environmental risk liability is material, financially speaking.

Regardless of the concern, the financial liability to the company arises from the peril or actual and perceived human and/or ecological risk described above and how it is defined. Figure 2.4 summarizes the inputs to the assessment of financial risks associated with environmentally impaired (contaminated) property. Again, note how the definition of the actual human and/or ecological risk has the potential to drive the financial risks upward through increased legal liabilities, engineering costs, and lower property value. While Appendix B provides detailed information, let us now look briefly at each of these financial concerns.

a. Management Problem

Dealing with environmental risks can be a cost of doing business to some, but to most corporate managers it is just a costly headache. The time and costs involved are only negative returns on investment:

- The costs of performing environmental assessment, investigation, remediation, and monitoring.

- Once environmental liability issues are monetized as material costs, the company needs to devote timely attention to the following matters:
  - shareholders need to be informed;
  - management needs to react, *i.e.*, take appropriate action to mitigate the liability;
  - sometimes funding reserves must be established and other appropriate corrective steps taken;
  - costs associated with carrying a non-performing, impaired property asset;
  - risk transfer costs;

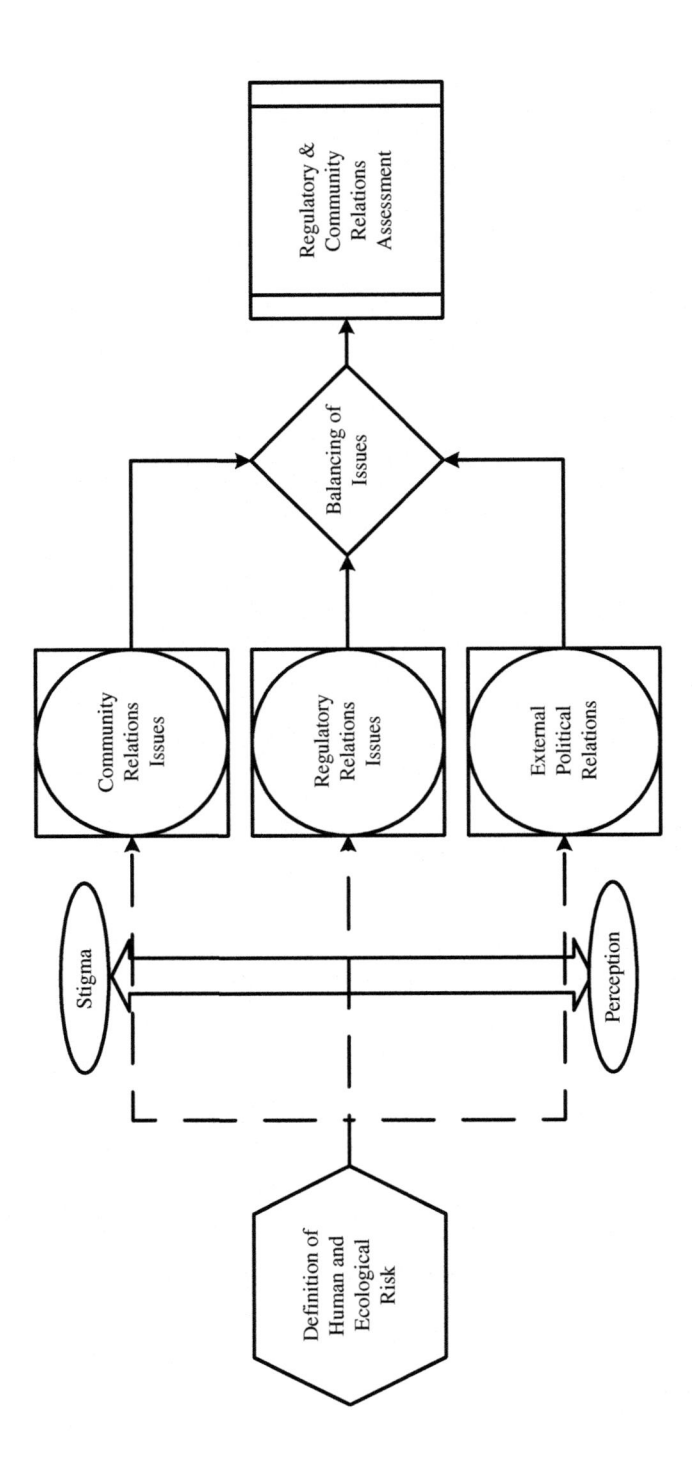

Figure 2.3 The definition of human and ecological risk as a peril and cause of loss drives concern within the company about its image and external relationships.

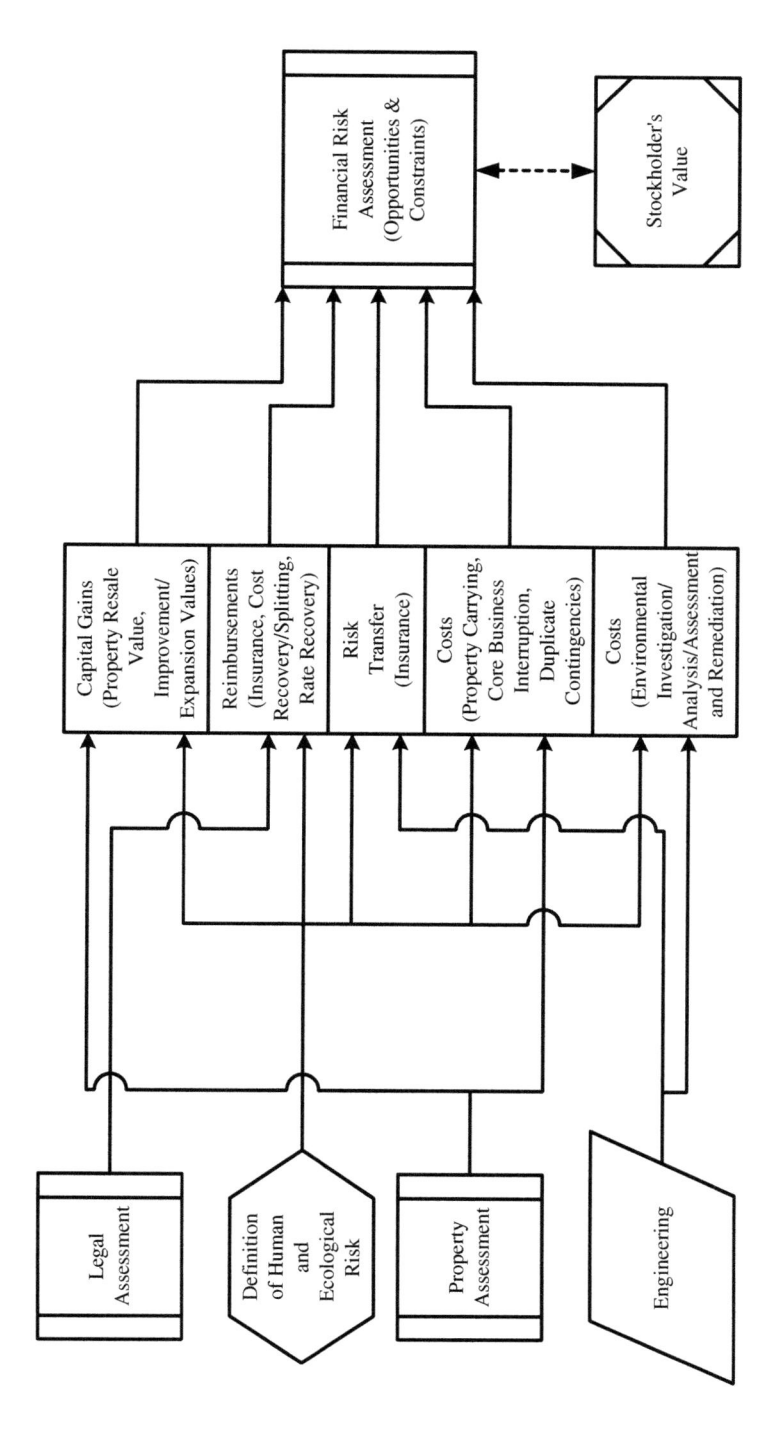

Figure 2.4 The definition of human and ecological risks as a peril and cause of loss to the company are a major contingent liability. This critical financial concern has implications for a variety of costs and other implications.

      ○   effect on capital gains via property resale value or improvement/ expansion value; and

      ○   opportunities for reimbursements from historical insurance policies, cost recovery or splitting options, and rate recovery potential.

- Additionally, in real estate transactions, the environmental risk issue becomes a parameter within the larger deal negotiations.

b. Detrimental Condition to Property Value

There are ten identifiable classes of detrimental conditions to real estate (p. 17 of Bell, 1999). One of these conditions is relevant to this discussion, namely Environmental Conditions (or Class VIII, see Exhibit 0.9 of Bell, 1999). These conditions include soil contamination, building contamination, naturally occurring conditions such as radon, and impacts to air and groundwater. Of course, actions can correct or mitigate these conditions. However, the costs associated with assessing, repairing (remedial or corrective action), and possible ongoing operational and maintenance costs have important financial implications to the owner, purchaser, tenant, and/or financial institution (if a lending transaction is involved). Figure 2.5 summarizes the property appraisal process and how detrimental conditions, such as environmental contamination, are factored into the definition of property value. Again, note how the definition of the actual human and/or ecological risk has the potential to drive the value down through negative market stigma and cost.

c. Impairment Limiting Site Use and Increasing Business Cost

The environmental impairment due to contamination may limit use of part or all of the real estate in ongoing operations or affect property transactions (again, due to cleanup liability and compliance liability). The financial aspects of cleanup liability include reimbursement for response and/or oversight costs of federal and/or state agencies, natural resource damages, and third-party claims.

d. Is the Liability Material?

As discussed in Appendix B, the principal financial issue in managing environmental liabilities is whether the aggregate cost of dealing with an environmental risk is material to shareholders. If so, the question arises whether the difference between the company's current monetary reserves and the expected liability cost is material to shareholders (that is, will it affect—decrease—the value of the stock, see Figure 2.4). The measure or definition of *materiality* comes from an unspecified percentage (generally from 3% to 10%) of total liabilities or operating income. For example, a company with $1 billion in total liabilities should consider funding an environmental reserve component if the environmental liability is in the range of $30 million to $100 million. If the same company has a quarterly operating profit of $100 million, a reserve increase of $3 million to $10 million would be material to that company's quarterly earnings.

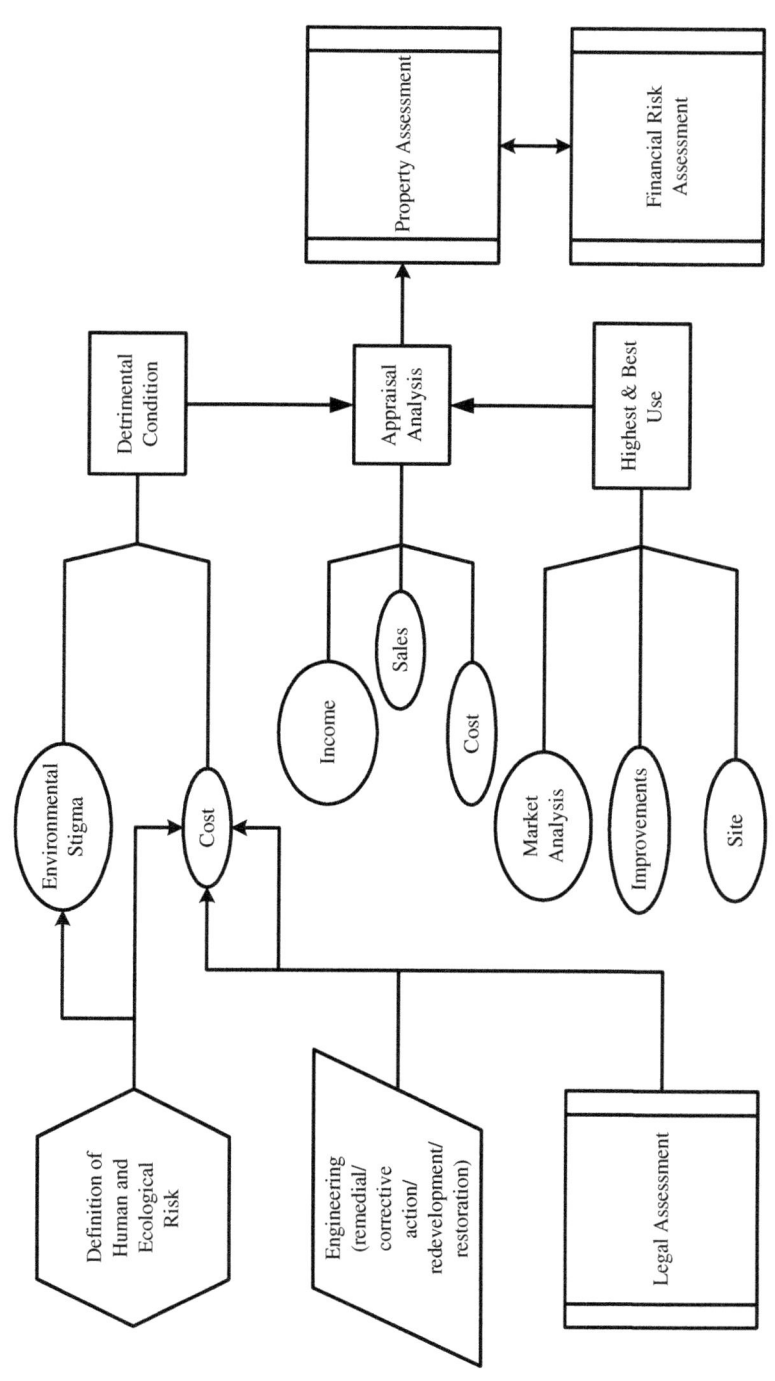

Figure 2.5 The definition of human and ecological risk as a peril and cause of loss to the company is critical in its effect upon the value of the company's property asset through the definition of a detrimental condition.

It may not be possible to realize or calculate every suspected environmental risk in financial terms today (for example, the impact of a company's products on biodiversity or global warming). Nevertheless, there are several statutory requirements for recognizing environmental risks as financial risks.

The conversion of environmental risks into financial risks occurs through a continuous screening and funneling of financial data to shareholders. The first step, *recognition*, comes from Financial Accounting Standards Board Statement 5 (or FASB #5). This statement requires recognition of a contingent liability when the loss is both probable and reasonably estimable. If the liability meets these requirements, then the company must estimate, using current information, the liability. The estimate may be updated later, up or down, but the cost has to be quantified at least to a defined range. If there is no best estimate within the stated range, then the lower end of the range is used as the estimate. Once estimated, the company sums all individual environmental liability costs together and subsequently reports the findings. The company uses the reported findings to test if the current environmental reserves match the current liability estimate. If the current reserve balance is material (see definition above), or if the company adjusts the reserve balance (such that quarterly earnings are materially affected), then the company must disclose that information to shareholders. This reporting may be part of the annual or quarterly financial statement process.

B. EVOLUTION OF THE REGULATORY RISK ASSESSMENT/ MANAGEMENT/COMMUNICATION PARADIGM

As discussed by Kervern (1995), before 1755 the solution to "catastrophes" came out of "magic." An earthquake in Lisbon, Portugal in 1755 inspired Jean-Jacques Rousseau to conclude that the tragic loss of life arose from faulty human decisions. According to Kervern, Rousseau's comment stimulated great debate laying the groundwork for a science of hazards (in the insurance context that is those conditions creating or increasing the chance of a loss).

The rise of insurance use led to the measuring of risks, for example, for the loss of life or the likelihood of flooding. In the mid-1900s, the advancement of technology stimulated the use of scientific methods in predicting accidents, especially those in the chemical process and nuclear industries. At about the same time (the late 1950s), the United States Food and Drug Administration (FDA) used risk analysis methods to evaluate human health effects from additives contained in food, drugs, and cosmetics.

Environmental statutes generally require decisions about what is "safe." USEPA in its first few years tried to describe and evaluate environmental status and trends, but how does one define environmental quality? This dilemma stimulated the focus on defining risks to the environment and people (Schierow, 1994).

Before 1970, no one really applied risk analysis methods to complex issues such as environmental hazards. USEPA and the Occupational Safety and Health Administration (OSHA) began developing new procedures and adapted existing ones. Each agency developed its own procedures to assess risk, crafted risk management policies, and made decisions to meet its "individual" perspectives. In 1977, the Interagency Regulatory Liaison Group (IRLG, and composed of USEPA,

OSHA, the Consumer Product Safety Commission, FDA, and USDA) responded to criticisms about differences in approaches, assumptions, etc., and particularly about cancer risk assessments (NRC, 1983). The IRLG proposed a policy in 1979 for coordinating the analysis and management of such risks across the agencies. Various concerns and controversy over this policy stimulated Congress to authorize the landmark 1983 study by the NAS on ways to improve the use of risk analysis within the federal government (see Appendix A for details).

Simultaneously, USEPA began proposing interim guidelines for its own assessments to address criticisms and calls for consistency. In response to the requirements of Superfund legislation (CERCLA), USEPA began publishing a series of documents for guiding agency practice. First, there was the draft Manual for Performing Endangerment Assessments published in 1984. The White House Office of Science and Technology Policy (OSTP) adopted the NAS framework in 1985, providing a basis for developing policy guidelines. In 1986, USEPA established final guidelines for analyzing risks of cancer and other health effects, all based on the NAS framework (51 *Federal Register* 33992-34054, Sept. 24, 1986). These guidelines addressed developmental risks, human exposure to individual chemicals, and risks from chemical mixtures. At the same time, USEPA published the Superfund Public Health Evaluation Manual or SPHEM (USEPA, 1986c), which supplanted the use of endangerment assessments. In 1987, a series of formal Risk Assessment Guidelines (called RAGs) began to be published. The Agency subsequently revised these guidelines and developed others.

In August 1994, an interagency work group for the Clinton Administration released *Draft Principles for Risk Assessment, Management, and Communication* to serve as a "general policy framework" for regulatory implementation. Following these principles, USEPA continues to update and extend RAGs providing a framework for ecological risk analysis, standardized methods of presenting risk data, improved risk characterization, enhanced risk communication, cumulative risk issues, and quantitative uncertainty analysis. They have as well developed guidance for addressing neurotoxicity and reproductive risks and improving exposure measurements. (See Appendix A for more information.)

So, where do we currently stand in the evolution of the risk assessment–management–communication paradigm? In the past 20 years, scores of publications on the subject from the federal government, interagency task groups, scientific consortia, and others have built upon the works of their like-minded predecessors to establish a policy model for risk analysis. That model still reflects strongly its origins in the paradigm of the NAS "Red Book." Regardless of the fine-tuning, the result is the same:

- A qualitative and quantitative estimate of risk for adverse effects to human health and the environment.

- The regulatory agencies (who define themselves as "the" risk manager) review and balance the results of a risk assessment and other lines of information. This review and balancing process is risk management: the evaluation and selection of alternative regulatory and non-regulatory responses to environmental risks, through the consideration of legal, economic, and behavioral factors.

- This review and balancing process requires interaction with the regulated community, the concerned public, and/or other interested stakeholders. The process then proceeds in time series:
  - o internal and external discussions and hearings;
  - o drafting of decisions;
  - o allowance of comments and critiques by affected and interested parties;
  - o response to all comments;
  - o publishing the final decision; and
  - o implementation of the decision.

Since its publication in 1983, however, focus has gradually shifted from a particular point of emphasis in the "Red Book." That errant emphasis was a strict conceptual separation between risk assessment and risk management, with the result being a practical insulary between risk assessors and risk managers. Fortunately, the shift in focus seeks to establish the process of analysis and management as a more interactive, even democratic, one, greatly encouraging multi-lateral communications (see Appendix A).

C. THE CURRENT "RISK-BASED" MOVEMENT

The concept of managing environmental impairment based on the amount of risk (likelihood and intensity) to human health and the environment is demonstrated throughout the history of environmental issues over the last few decades. In the 1980s, with the realization of the massive size of dealing with the petroleum underground storage tank (UST) issue, many regulatory agencies simply applied standards developed for other purposes uniformly to UST release sites to establish cleanup requirements. It quickly became apparent to those in industry that such standards, without consideration of the extent of actual or potential human and environmental exposure, led to an inefficient, costly, and time-consuming process, all of which is anathema to corporate managers. Although various risk analysis, management, and decision methods were available, industry sought to streamline and standardize the process associated with site assessment, standard-setting, and corrective action as it applied to petroleum-impaired properties. Driven by their belief and empirical confidence in science-based risk analysis (as well as a focus on excluding "subjective" bias), they developed "risk-based corrective action" or RBCA (commonly pronounced as "Rebecca"). They firmly believe, and have convinced many others, that this is the system to assure protection of human health and the environment. Its success is demonstrated by its evolution in the standards setting process of the American Society for Testing and Materials (ASTM), acceptance by USEPA, and use as a model by many state environmental agencies.

Even so, an underlying concern within industry continues—particularly with the increased use of stakeholder processes—that, without a strong and consistent scientific basis, subjective issues and approaches will control risk management processes (Charnley, 2000). Should this happen, many in industry believe that it will lead to incorrect priorities and/or ineffective expenditures (*i.e.,* spending a great deal of time and money on concerns that constitute little in the way of a significant peril). There is evidence that this does indeed happen (Ames and Gold, 2000), but Charnley (2000) presents several successful case studies where these difficulties were overcome. Unfortunately, these instances were notably large and complex situations and not the more typical circumstances associated with smaller impaired properties dealt with by most managers.

D. THE PRECAUTIONARY PRINCIPLE

The precautionary principle requires the taking of action to reduce the potential for adverse effects on human health and the environment from chemicals, products, or processes prior to establishing scientific proof of harm. Sandin (1999) points out that at least nineteen formulations of precautionary principle exist, differing on many levels. The Rio Declaration (1992), which perhaps is the one use of the principle with the widest support, states:

> *"In order to protect the environment, the precautionary approach shall be...applied.... Where there are threats of serious or irreversible damage, lack of full scientific uncertainty shall not be used as a reason for postponing cost-effective measures to prevent environmental damage."*

At a conference held in January 1998, thirty-two environmental experts from around the world discussed the role of the precautionary principle in regulations and environmental management, and agreed that

> *"...new principles for conducting human activities are necessary.... When an activity raises threats to the environment or human health, precautionary measures should be taken, even if some cause-and-effect relationships are not fully established scientifically. ...the proponent...rather than the public, should bear the burden of proof...."* (Hileman, 1998)

Proponents of the principle claim that it allows the regulatory community to cope with true uncertainty. Risk assessment is a tool used to estimate statistical uncertainty. However "true" uncertainty or indeterminacy, *e.g.*, what your risk of developing cancer is if you live near a disposal site for some newly developed toxic chemical, has yet to be adequately addressed in environmental protection strategies (Costanza and Cornwell, 1992). The majority of promulgated environmental protection regulations, particularly in the United States, focus on cleaning up or containing contaminated media, not the prevention of contamination. Advocates say a shift in focus would occur with the adoption of the precautionary principle.

Although the principle has enjoyed some success in the United States and gained acceptance internationally, critics argue that as a matter of practical application, the principle falls short. Concerns about implementation of the principle fall into two major categories.

1. The principle does not specify the situations in which precautionary action is required, nor does it identify what precautionary measures to take.

2. The idea that the producer should bear the fiscal responsibility for adverse effects on the environment is often not practical. This impracticability

arises because it is difficult to determine how much the producer should pay or when.

In short, the principle does not include guidance regarding the financial commitment required to safeguard against potentially adverse future environmental or health situations. These issues hinge on the theoretical efficiency to be achieved in environmental management via market mechanisms (Costanza and Cornwell, 1992).

The principle has other deficiencies, perhaps the greatest being the variability of its interpretation (Vanderzwaag, 1999). The principle also does not clarify how to use formal uncertainty analysis to achieve precautionary desires, nor is the role and application of science and the ability to use technology in its implementation straightforward (Graham, 2000). Thus, there is no clear way to reconcile the principle with the application of risk analysis and its supporting base of science (Foster *et al.*, 2000). Although international law and certain countries incorporate the principle, much work remains on how to incorporate science and the balancing of available information within a politically transparent process.

Even so, some feel that the efforts of using science and risk assessment, in its quantitative form, over the last couple of decades have driven environmental regulators away from *protection* of public health and the environment. Vincent (1999) concludes that,

"All risk assessments are wrong, some are useful."

From this perspective, Vincent does not see the decision process of environmental management as an objective, scientific process, but as a value-laden political process. He sees the issue "for people at risk" as one of precautionary prevention, not risk. From our experience as risk assessors and environmental consultants, many stakeholders voice the same sentiment. From the manager's perspective, how does one cope with this view?

III. THE MANAGEMENT OF ENVIRONMENTAL RISK

So far in this chapter, we have summarized how various parties approach the matter of defining environmental risk—in a company, by regulators, by those who prefer science, and by those who favor a precautionary perspective. The current evolutionary state of the regulatory paradigm for risk, which drives most of the decision-making on environmental issues, is incomplete. This is because the paradigm, originally crafted for regulatory policy purposes, fails to assist today's corporate environmental manager in facing the bipolar reality we already have observed:

- Advocates of risk-based decision-making on the one hand, who champion science and consistency of approach, but at the same time appear to critics to be no more than champions of less cleanup.

- Those who raise their voices for "precaution" who, out of simple concern or fear of the unknown and/or the impossible-to-ever-know, cry out that there can never be enough cleanup.

Environmental risk management is a process of managing business risks, such as property loss and liability, but it is much more.

A. CLASSICAL RISK MANAGEMENT

As shown in Figure 2.6, risk management as a classic business process involves several steps (Neuman, 1998):

- **Identification and analysis of loss exposures.** Loss exposures are the possibility of a financial loss due to some peril striking something of value, such as property, income, liability, and personnel. This step requires managers to spend time identifying loss exposures. Once identified, characterize the loss exposure in terms of its source, likelihood, intensity, effects, and the potential for avoidance or, if unavoidable, mitigation.

- **Evaluation and selection of alternative risk management techniques.** There are two approaches:
  - ○ Risk Control (avoidance, prevention, reduction, segregation, or contractual transfer), and
  - ○ Risk Financing (either through retention, such as reserves, or transfer, using insurance or contractual mechanisms).

- **Risk management administration** (which includes management implementation and monitoring of effectiveness).

From a general business perspective, the most important risks are those that are severe, occur frequently, and are cheap to fix (Grose, 1987). Such risks rank at the pinnacle of Grose's "Hazard Totem Pole." This ranking system considers risks on a tripartite scale, with each dimension weighted on the manager's perception of importance:

- severity (how much would they affect normal corporate performance);

- probability (how likely are they to occur); and

- potential loss (the cost of the loss versus the cost to fix the problem before a loss is realized).

However, as discussed earlier, a property with an environmental impairment has a peril with more dimensions and people involved than most loss exposures to a company. It is more complex and difficult to manage because of these things.

B. ENVIRONMENTAL RISK MANAGEMENT

Environmental risk management is similar to the classic process with at least two important exceptions as shown in Figure 2.6 and described by Neuman (1998):

- As discussed earlier in this chapter, environmental liabilities involve legal and financial issues that are more complex than other liabilities. This complexity requires careful problem formulation to identify and evaluate the nature of the environmental issues involved using environmental and risk assessments.

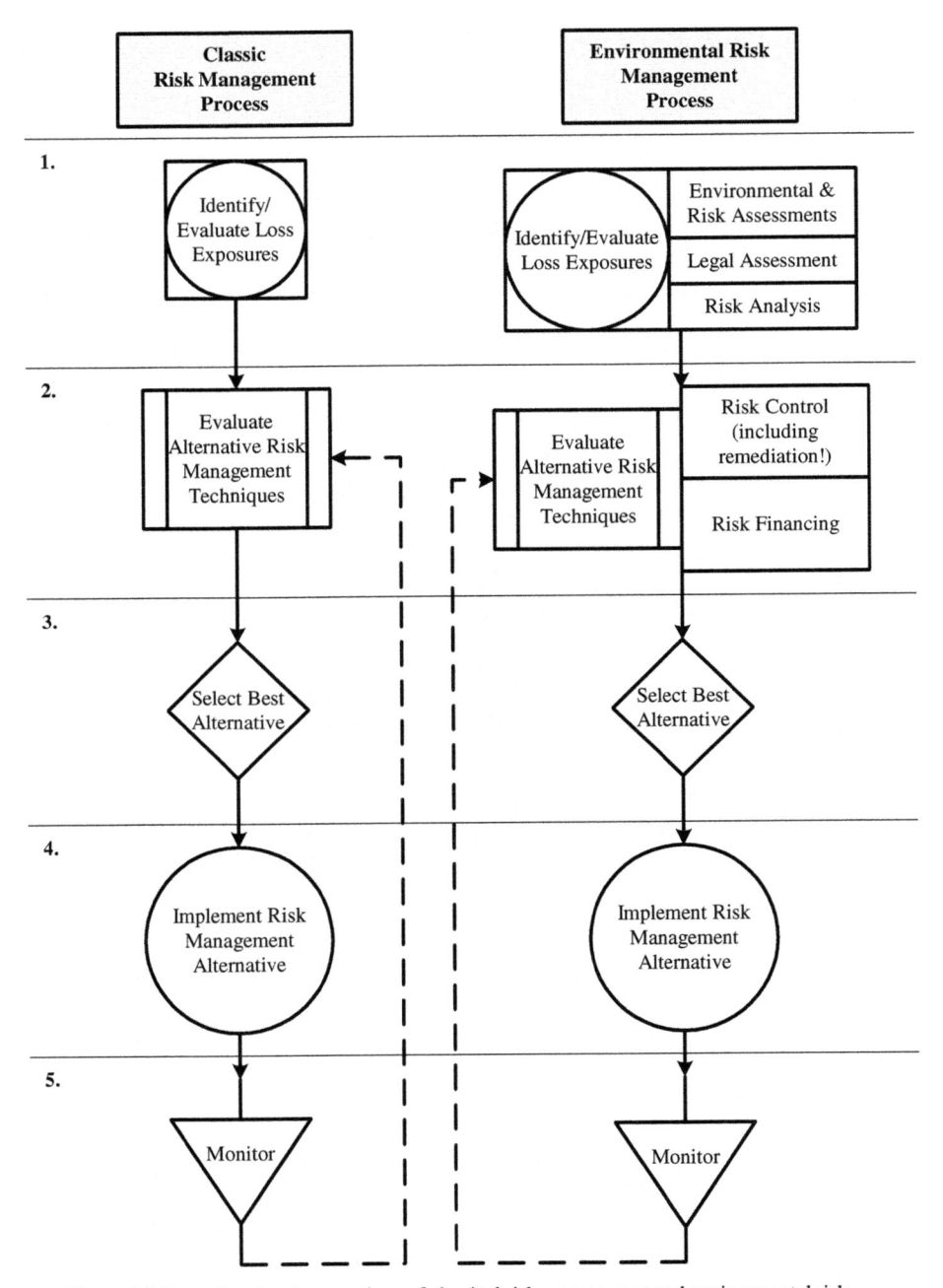

Figure 2.6 Cross-functional comparison of classical risk management and environmental risk management processes.

- These assessments need to provide data, information, and insight about the physical, chemical, and biological science basis, the engineering basis, and the sociopolitical/demographic basis of the environmental issues. It is equally important to assess the legal issues (local, state, and federal requirements and programs). All of this must then be combined through a profiling technique to appraise the business or financial risk (that is, quantify the loss exposure in financial terms).

- As Neuman (1998) indicates, remediation is the foremost risk control technique, and environmental departments essentially dominate the process as compared to risk management departments. However, risk financing now is more widely available and cost-effective, but this does not necessarily mean that they are a substitute for remedial action or some kind of institutional control.

Environmental risk management also is more complicated because of the decision-making aspects involved (Chechile, 1991):

1. Their sheer complexity and scope.

2. We do not know as much as we like to think or as we actually need.

3. Dynamic systems over space and time.

4. Public involvement and political pressures.

5. Environmental economics and related market force dynamics.

6. Ethical issues arising from the interaction of individual and corporate.

7. Environmental issues are trans-boundary in nature.

Now, is the business risk of environmental liability unlimited cleanup requirements, uncertainty about the time horizon involved, or is it just the money? The business risk is primarily due to unquantified scenarios, in terms of potential remedial (risk control) activities based on the property's current use, as well as its highest and best reuse prospects (Ackerman and Soler, 2000). Such scenarios result in unquantified requirements to control (remediate) the impairment, which act as the peril producing risk.

Ackerman and Soler (2000) further indicate that it is meaningful for the corporate manager to think about environmental risks as a function of the time required for remediation and the money involved. Once sites have business parameters that are defined and consistent with the corporation's purpose, including a defined time horizon, then (internal and/or external) financing can be arranged and appropriate risk transfer techniques (*e.g.*, stop loss, cost cap of remediation, and pollution prevention insurance) implemented, together with the necessary engineering steps. An environmentally impaired property with no known data set or set of quantifiable business parameters cannot have a timeline projected nor financing structure placed. Most critical here is the challenge of remediation standards ("how clean is clean?"), their consistency of application, and obtaining mutual agreement among interested parties, especially the community surrounding the impaired property. Without definition of the applicable standards, clarification of divergent objectives, and mutual agreements on the planned result, remediation remains unquantified and the associated liability risks are difficult to place in understandable financial and legal contexts.

C.   THE PROBLEM: MANAGING HOW RISK IS DEFINED

The complexity of environmental risk management and the difficulties of setting standards and answering the question "how clean is clean," appear—at least to us— to hinge on a single management problem composed of three questions:

- How to define the peril (that is, the human and ecological risk)?

- Who defines the peril?

- When is the definition of peril made?

It is axiomatic that risks to human health and the environment—the peril— cannot be absolutely quantified. As Kervern (1995) states, any risk measurement is relative to the person making it and subject to observational constraints. It is impossible to prove that anything is entirely harmless, let alone an environmental impairment due to "chemical contamination" (Huning, 2000). The matter has political dimensions (Vincent, 1999). Thus, risk measurements in this context depend upon socially set rules and values (conventions) pertaining to the issue at hand; even so, it must be appreciative of sciences' value and limits. From this perspective, it is understandable why the evolution of the regulatory risk paradigm of assessment/management/communication has begun to include more stakeholder processes in the last decade.

We began this chapter discussing the trans-boundary nature of environmental risk management, but working across boundaries means that any involved processes require interfacing, especially when dealing with different gestalts. Environmental risk management is dependent upon an organization's goals, and these goals must be sensitive to the power of scientific information and its limits to knowing. This management process must also appreciate as well the extant sociopolitical milieu in which any decision will be wrought. Thus, it is critical for the corporate manager to develop and maintain a multidimensional view of the meaning of environmental impairment, in order to understand the concept of and effectively manage risk.

IV.   A MULTIDIMENSIONAL VIEW OF ENVIRONMENTAL IMPAIRMENT AND THE ASSOCIATED RISKS

To be effective strategically, environmental risk management requires consideration of not just the hard, objective, scientific facts but thoughtful consideration of the subjective and even political nature of these issues. To be effective tactically, the manager must interface both into a synoptic understanding, which leads to a defined management plan.

A.   VIEWING RISK FROM MULTIPLE PERSPECTIVES
1.   Risk and Rationality

In her 1991 book *Risk and Rationality*, Kristen Schrader-Frechette argues for a theory of rationality operating within social decision-making as it pertains to risks to health, the environment, and general well-being. Her approach of *scientific proceduralism* steers a moderate course between:

- cultural relativism (where proponents conclude that risk is a collective, social construct and that more often than not a "citizen's" assessment is biased) and

- naïve positivism (where the experts assume that measurements of risk can be valueless and objective, and a "citizen's" assessment is purely irrational).

Scientific proceduralism seeks to apply science within a more democratic process. Although she argues against the usual approaches to quantitative risk assessment, risk evaluation, and risk-cost benefit analysis, Schrader-Frechette strongly urges their continued use as analytic informational tools. Through a more democratic process, she hopes that citizens can become more involved, informed, and capable of providing consent through negotiations. More importantly, whether one agrees with her or not, Schrader-Frechette makes a powerful case that lay conclusions about risk are more rational than most experts (or business people) believe.

2. Framing

As discussed in Chapter 1, conceptual models are crucial tools for bounding (or framing) those segments of the environment associated with a particular property asset with which one is dealing. Their critical feature is simplifying the spatial and temporal characteristics of the management problem facing the corporate manager. They not only aid to our focusing but also present how we perceive components of the system and the relationships among those components. These models thus provide a physical description of the factual and theoretical aspects of the manager's frame of reference (Rein, 1983 and Swaffield, 1998). They also present crucial insight into the manager's values and actions. The value of defining an environmental risk management problem with such models is that it serves to explain the structure by which we understand the problem, helps us identify the objectives we seek to obtain, and helps us demonstrate the methods for arriving at certain judgments about the problem or issues presented.

However, these models are usually prepared by (paid) technical experts (consultants) trained in science to make observations and measurements of various parameters. Those experts combine these parameters into an interpretation or explanation, confirming them through comparisons with other data and knowledge sets and through techniques of formal validation and peer review (internally and externally). There are four important points to remember about such models:

- The models of experts are not objective descriptions of the world, in that they suffer from generalization and distortion arising within the expertise of their own socially constructed belief system (Wynne, 1984).

- As simplifications, the models are dynamic and subject to change, as more information becomes available.

- Based in science, expert's models are falsifiable, but they are not provable as being "correct."

- Frames of reference and their associated models are "operationally sticky" in that people generally start from what they know and change existing ideas, or models, to help them understand and cope with new problems and issues. Also, there are often institutional, economic, and practical considerations tied to the use of models. These considerations mean that decision-makers (the corporate manager) and modelers (consultants) are given incentives to use and sell existing models regardless of their appropriateness (Wynne, 1984).

Regardless of their weaknesses, models are beneficial and powerful in building an understanding of the problem and decisions at hand—that is, as long as people acknowledge their sources, inputs, assumptions, and limitations, and can articulate to themselves and others how their own values and politics intersect (Tong, 1986 and Freudenburg, 1988). Moreover, conceptual models and framing can help capture the reference frames of other stakeholders.

3. Shared Understanding and Legitimization

How does one deal with a multiplicity of frames of reference concerning the objective and subjective data, information, values, and opinions about risks to human health and the environment? Menzie (1998) suggests the importance of communications in resolving alternative worldviews based in the seemingly counter-current or oppositional frames of reference of perception-based reality and science-based reality of health and ecological risk issues. The key to this communications effort is the development of a *shared understanding*.

We have discussed the development of expert knowledge and its role, but what role should lay knowledge play in managing environmental issues, particularly issues involving the property assets of a corporation? According to Lopez Cerezo (1999), adequate environmental policies, or decisions pertaining to an environmental impairment upon your property within some local community, need to:

- Be efficient and legitimate.
  - o Efficiency in this context means that the policy uses sufficient resources to achieve the stated objectives of a particular agenda.
  - o Legitimacy means that the policy has social support through either positive public perception or explicit democratic support.
- Emphasize the use of pragmatic knowledge.
  - o This type of knowledge is in contrast to realist (or purely objective) knowledge.
  - o It involves claims or assertions that are warranted or justified within a particular social, cultural, and historical context.

These requirements suggest that expert knowledge has to be negotiated as acceptable, and this acceptability is judged within two tribunals:

- The tribunal of nature—here the data, information, understanding, and knowledge offered by experts are judged as scientifically sound, generally by peers.

- The tribunal of society—here the same data, information, understanding, and knowledge are *rendered* politically legitimate, generally by lay people.

Such a process suggests the need for the appropriate inclusion of lay knowledge to identify and counterbalance expert indeterminacy and uncertainty. In so doing, the process remedies expert biases and contributes to mutually acceptable solutions. Perhaps, as Lopez Cerezo argues, political legitimacy is essentially equivalent with technical correctness.

Are there formal processes that corporate managers might use to achieve shared understanding and legitimization? Appendix A describes a couple of important developments along this line. Specifically, *Understanding Risk: Informing Decisions in a Democratic Society* (NRC, 1996) and the Presidential/ Congressional Commission on Risk Assessment and Risk Management that published the *Framework for Environmental Health Risk Management* (PCCRARM, 1997a). However, before these approaches, Morgan, Fischhoff, Bostrom, Lave, and Atman (1992) advocated a four-step procedure to develop a risk communication approach. Their approach is an empirical exploration and validation process. While robust and highly capable, this procedure is practically unusable for most corporate managers. Nevertheless, it is highly informative to this discussion of risk, rationality, and problem-framing in that it points out important things to remember when seeking to manage the definition of environmental risk. The four-step risk communication procedure of Morgan *et al.* involves:

- An open-ended elicitation of beliefs people have about a hazard.

- Deduce the prevalence of different beliefs using structured questionnaires.

- Develop a communications approach to meet people's needs based upon a psychological assessment of their current beliefs.

- Test the communications approach and refine it as necessary.

Essential to this approach is the building of a model about how a layperson perceives the particular environmental issue. This is a "mental model" of how a person processes information pertaining to the various factors relevant to a hazard-related decision. Built as an Influence Diagram, the model serves as an organizing device (like a decision tree) presenting a directed network showing relationships between relevant factors that lead to and influence a particular decision or set of decisions. This model thus provides a template for characterizing a person's mental model (or frame of reference) summarized as follows:

1. Statement of belief.

2. Appropriateness of the belief (that is, is it accurate, erroneous, peripheral [correct but irrelevant], or indiscriminate [to imprecise to be evaluated]).

3. Specificity of that belief (how detailed is it).

4. The category of knowledge associated with the belief:
   ○ exposure processes,
   ○ effects processes (that is, health, ecology, etc.),
   ○ mitigation behaviors (can it be fixed or avoided),
   ○ evaluative beliefs (such as, "it's bad"), and
   ○ background knowledge (about the area, timing, and substances involved).

The value of the Morgan *et al.* approach to communicating risk to the public is its emphasis on learning what people know and believe about the issue and decision process at hand. This emphasis is a part of the Risk-Based Analysis approach (called Situation Analysis) discussed in this book (see Chapter 4 §III).

B.   INTEGRATING A BIPOLAR ISSUE

In 1995, Kervern described in a short communication a managerial approach to risk management called the science of danger or cindynics. Using this theoretical approach, a manager can systematically work through the various dimensions or perspectives that influence a situation, which results in a risk threatening a company's financial and physical assets. Kervern describes several rudiments to understanding danger or the risk of hazardous situations and the measurement of this risk:

- Danger or the risks of a hazardous situation is/are relative to the observers, based on the conventions established by the human network concerned with the issue(s), and dependent on the goals set (by that human network) for the measurement process itself.

- The measurement of risk is ambiguous due to the interaction of five operational areas (Figure 2.7):
  o   problems in defining the goals and objectives of the measurement;
  o   problems in defining the models of risk used in the measurement;
  o   problems associated with the data, statistics, and knowledge used in the models of risk;
  o   conflicts concerning rules governing operations within the human network where the danger or risk (is feared to) occurs and is measured; and
  o   value systems operating within the human network.

- Reduction in ambiguity comes through investigation.

- Crisis is a destruction of human knowledge networks; thus, crisis management involves creation of substitute networks for the destroyed ones.

- Human influence causes danger to wax and wane.

These rudiments, especially the five operational areas effective in risk measurement, are very important in this current discussion. Figure 2.7 depicts these dimensions or domains. It should be obvious that each stakeholder will have her or his own set of rules and values. They will appreciate, to some degree, the goal and objectives of the measurement effort. Additionally, they will have opinions on the models used to measure risk, as well as the data, statistics, and knowledge used to produce that measurement. Just as important, and perhaps even more important, are the "interactions" or overlaps among the operational areas (or "spaces" to use Kervern's words), which result in six domains:

- Goals/Objectives and Models of Risk: create the domain of what is *practical* in terms of measuring risk.

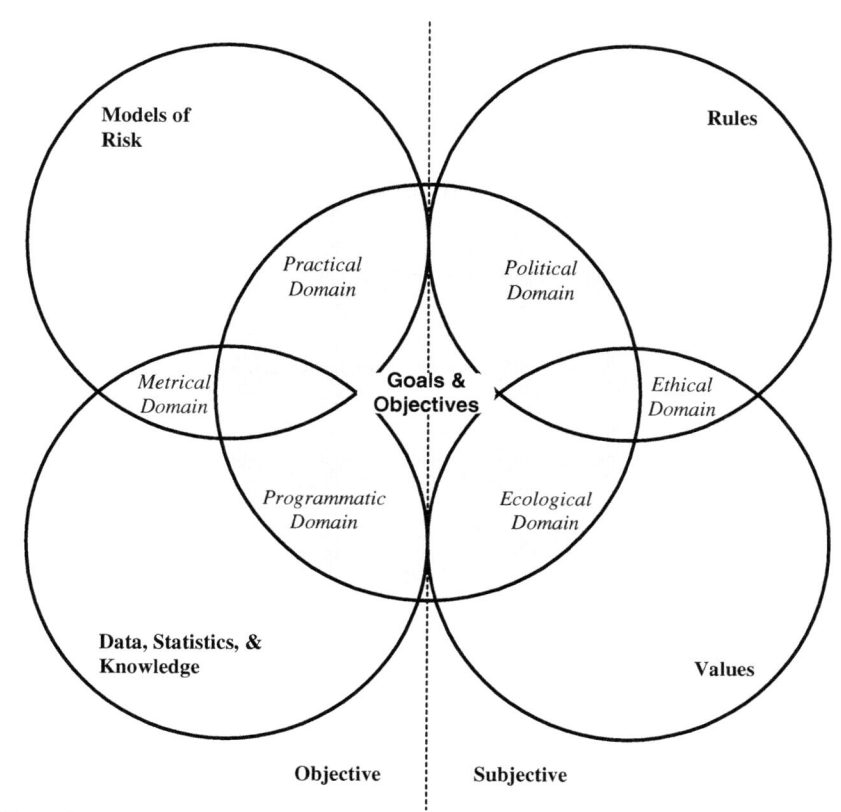

Figure 2.7 The multi-dimensional nature of risk and the operational domains affecting its measurement, interpretation, and management as suggested from cindynics.

- Goals/Objectives and Data/Knowledge: create the domain of ***programmatic*** requirements for measuring risk.

- Models of Risk and Data/Knowledge: create the domain of the actual tools (***metrics***) for measuring risk.

- Goals/Objectives and Rules: create the domain of what is ***politically acceptable*** in terms of measuring risk.

- Goals/Objectives and Values: create the domain of ***ecological*** aspects, including "quality of life," considered in the measurement of risk.

- Rules and Values: create the domain of what is ***ethical*** in terms of measuring risk.

Classifying these six domains into two types, we see that the management of environmental risk is fundamentally bipolar:

- The practical, metrical, and programmatic domains are *objective* aspects of the measurement of risk.

- The ecological, political, and ethical domains are *subjective* aspects of the measurement of risk.

While these aspects are distinct and differ in their dynamics, they are indispensable to our understanding of a risk system because their mutual relationship defines one another. There is a great intellectual temptation for oversimplification, emphasizing either one and marginalizing or eliminating the other. The challenge is to recognize and integrate both aspects, and to address and effectively balance each domain. It is generally easy to understand the dimensions of time and space of a risk system (Figure 1.2). It also is generally straightforward to cope with the regulatory issues (programmatic goals and objectives) and the data and information, which fill the scientific and engineering models used to measure risk. However, appreciating the subjective nature of risk and building a management approach to address the human networks involved is as hard as it is valuable. The cindynic conceptual framework provides a structure for understanding the importance of effectively interfacing information, analysis, perception, and communication. Applying it will aid the corporate manager to define the environmental risk to be managed, which will lead to a more efficient and legitimate result while defusing tension and crisis.

V.   MANAGING THE DEFINITION OF THE RISK

In this chapter we have reviewed classical risk management and the differences inherent in dealing with environmental risk, discussed how risk is defined, and established that the subject is fundamentally multidimensional and transboundary. Let us now conclude this chapter by looking at environmental loss exposure in terms of hazards, risk, costs, and understanding four crucial qualifiers: risk importance, risk as harm, risk significance, and the uncertainty of risk. This précis will lead us to suggesting a holistic way that managers can approach dealing with this bipolar reality in their own operations, and thereby constructively manage the definition of environmental risk that they seek to manage.

A.   DEFINING LOSS EXPOSURE AS HAZARD, RISK, AND COST
Let us review again some appropriate definitions from earlier in the chapter:

- **Hazard**—a thing(s) or condition(s) that create(s) or increase(s) the chances of a business loss through the potential to cause harm (which means adverse outcomes to human health and ecological functioning including morbidity or mortality.)

- **Risk**—the chance aspect of the loss exposure, which has two components (Figure 2.1):
  - Actual Risk, which is the probability that a hazard will result in the occurrence of harm at some frequency and with some level of intensity, and
  - Perceived *Risk*, which is the importance and significance of a hazard and its cause of harm, as defined by individuals, the community, and society.

- **Cost**—which includes the cost to assess the risk issue and mitigate or fix (remediate) it over a defined time horizon, and the cost to manage the associated (see Appendix B):
  - ○ internal problems, legal liabilities, and image effects;
  - ○ detrimental condition causing damage to the property's financial value;
  - ○ use limitations of the site and/or increases to other business costs; and
  - ○ financial materiality of the environmental risk liability.

Defining the loss exposure associated with an impaired property is a matter of defining and assessing the peril or risk in terms of the various corporate functions or concerns identified earlier in this chapter:

$$Loss = f\,(health\,\&\,environ.\,concern) + f\,(legal) + f\,(image\,\&\,relations) + f\,(financial)$$

The corporate manager tasked with environmental risk management needs a process to manage these critical corporate functions effectively. However, there are mental barriers separating those who calculate an objective measurement of risk from the users of that measure, and these should be contrasted:

- Consultants, scientists, engineers, and risk assessors think about the objects they produce, *i.e.,* reports and plans, specifications, and construction.

- Users of these objects—corporate managers, regulators, and stakeholders— think about what the object(s) (that is, the many reports, their findings, conclusions, and any remedial action) do *for them,* or *to* them.

It is essential that this process thoughtfully interface various corporate functions with the definition and measurement of risk (Figure 2.8). The resulting balancing of issues and coping with the ever-present political realities within a corporation is a tough task for any manager. Nevertheless, this balancing act is essential to formulating a holistic strategy that addresses an environmentally impaired property with a synoptic understanding of environmental risk.

B. DEFINING RISK SYNOPTICALLY

Let us consider the strategic approach of using risk assessment in making decisions about environmental protection priorities. The approach described by Llewellyn (1998) is that of the Environmental Agency of England and Wales, which sought to encourage the use of the best science and analysis, and a synoptic view of the environment in decision-making. Their approach brings out the critical issues of importance, harm, significance, and uncertainty in managing the definition of risk. These issues are critical because of the weight of credibility (the legitimacy spoken of by Lopez Cerezo, 1999) required of decision-makers involved in environmental risk issues.

1. Risk Importance

As previously discussed, experience with environmental issues demonstrates that stakeholder groups have opinions that need acknowledgement and that need to be validated and accommodated, as appropriate, in an overall definition of risk. As Appendix C discusses, if the public and media say it is a crisis, then it is a crisis!

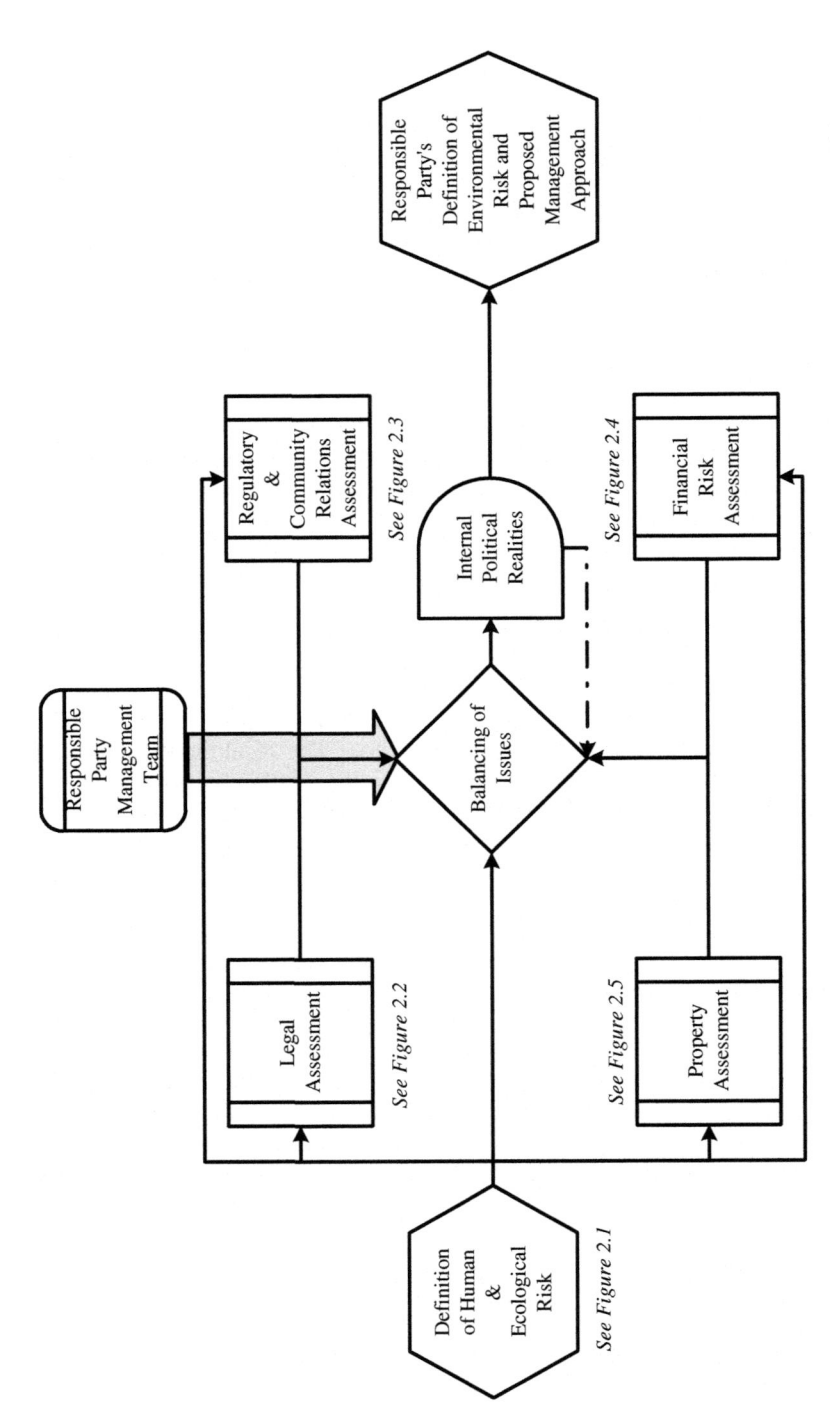

Figure 2.8 The definition of human health and ecological risk (the peril) as the cause of loss drives all of these components of the Environmental Risk Management Process. It is essential to interface all of the component assessments in order to formulate a holistic strategy for addressing the environmentally impaired property and to develop tactical approaches to legal, regulatory, community, financial, and environmental management issues that may arise during the follow-on implementation process.

There are problems with this conceptualization, however, as Peter Huber (1987) points out:

> *"Why [spend] time [on] risks that [are] either minuscule or...nonexistent...the answer [is] the sharp difference between the actuality of risk and its perception.... We fail to be frightened about things that should frighten us. This reality is...often embraced in...regulatory circles."*

Huber's point is that regulating toxins because of fear instead of what actually injures makes tests of importance mute or irrelevant. Scientists and corporate executives alike cite this fear (Charnley, 2000). However, not appreciating what is important or relevant to the community surrounding your impaired property can result in a credibility loss, outrage, and unstable or unworkable decisions (Susskind and Field, 1996 and Harris and Harper, 2001) and exacerbate economic inefficiencies (Lopez Cerezo, 1999). Harris and Harper suggest that it is vital to make inquiry to identify those categories of risk or impact that are important to stakeholders, such as:

- Public or human health effects, based on individual-, population-, and community-level metrics relevant to the lifestyles in the area, sensitive sub-populations, multi-generational concerns, and co-factor issues (such as multiple exposures and nutritional status, among others).

- Ecotoxicological effects, based on individual-, population-, and community-level metrics relevant to the impaired media and local attributes, critical habitat and resources, and co-factor issues (ecological co-stressors such as physical, thermal, and fragmentation, among others).

- Ethno-habitat effects such as those concerning natural goods, services, functions, and uses in the area.

- Socio-cultural effects to historical/cultural-resource elements/attributes and (native) lands access and use, for example.

- Socioeconomic effects.

These kinds of information provide site-specific insight about the multidimensional reality of the environment and its impairment. Such an approach follows the suggestion of Morgan *et al.* (1992) concerning the necessity of learning what people know and believe about an issue and decision process at hand. This relevant information helps the manager frame the definition of risk using multiple measures of risk and thereby balance the actual and objective with the perceived and subjective.

2. Risk as Harm

Harm is defined in the *American Heritage Dictionary* (Houghton Mifflin, 2000) as physical or psychological injury or damage, a wrong, or an evil. Earlier we defined harm as adverse outcomes to human health and ecological functioning, including morbidity or mortality. Along these same lines, harm in Llewellyn's (1998) context is the traditional scientific assessment of exposure and risk to a defined population(s) or critical habitat/natural resources and the evaluation of those results. "Risk as harm" is generally understood as a presentation of a basic risk system outside of its cultural contexts (as shown in Figure 1.2). Of paramount concern in the approach as discussed by Llewellyn is that the results of this analysis must be:

- "real" (with an understanding of the uncertainties involved),

- reasonable given the circumstances (that is, appropriate consideration is given to the assumptions made and the parameters used), and

- consistent with the regulatory policy context.

At this stage of the process, the measures of health or environmental risk may not necessarily fit within anyone's particular subjective frame (gestalt) regardless of its validity. This "risk as harm" is essentially the measurement generally required under federal and state regulatory guidelines in the U.S.:

- Risk or hazard for health effects (*i.e.*, public health risk)
  - Cancer—To characterize potential carcinogenic effects, intakes are combined with chemical-specific cancer potency slope factors resulting in a risk estimate. The estimate is the probability of that exposure resulting in an excess incidence of cancer, that is, the occurrence of more cancers than would normally be expected in that population.
  - Noncarcinogenic (or systemic health) effects—To characterize potential effects for substances that act systemically (including various vital organ systems, as well as developmental, neurological, and other effects), comparisons are made between projected exposure dose and reference doses. The comparison is made by calculating the ratio between the estimated (sub-)chronic daily dose to the corresponding reference dose. This ratio is called the Hazard Quotient.

- Ecotoxicological risks and hazards are estimated in a manner similar to systemic human health effects.
  - The total exposure is the sum of the organism's various exposures from diet, sediment ingestion, and surface water ingestion, for example:
    $$EE_{total} = EE_{diet} + EE_{soil} + EE_{water}$$
  - Doses of site-related chemicals are estimated for ecologically relevant and especially important or sensitive biota and compared to appropriate ecotoxicological benchmarks indicating threshold health effects.
  - Again, a comparison is made by calculating the ratio between the estimated dose (EE_{total}) to the corresponding ecotoxicological benchmark. This ratio is similarly called the Hazard Quotient or Hazard Index. By referring to the percentages of exposure resulting from different pathways (*e.g.*, food ingestion, soil ingestion, water ingestion), the relative contribution to total potential risk for each exposure pathway is identified.

However, if a manager develops an understanding of risk importance, then it follows that we need measures of "risk as harm" that not only include objective measures of adverse effects to public health and the environment, but also to the more subjective aspects mentioned previously. This is the lesson of the bipolar nature of risk as a danger as told us by cindynics (Kervern 1995).

The assessment or characterization of the risk as "harm" in terms of socio-cultural health (Harper and Harris, 2001) uses various measures, for example:

- Social/demographic indicators,

- Historic/archaeological/ religious/cultural resources and landscapes,

- Employment and economic growth, and

- Land use patterns and changes.

By way of example, the author led a team that developed a three-dimensional analysis of a risk system involving the pesticide DDT at the Bandelier National Monument in New Mexico (Ecology and Environment, 1996). This long-standing problem involved not only public health and ecological concerns, but implications to important historical and cultural aspects of the monument. This risk assessment considered the classical ecotoxicological and human health effects and evaluated the implications to the cultural/historical landscape, archaeological sites, and native and traditional cultural properties. The analysis aided the United States National Park Service in its discussion with state and federal environmental regulators and natural resource trustees in deciding what to do and how much. The analysis suggested that while the pesticide was present, it posed no current or future risks for humans (park rangers or visitors) or the important ecological resources present. Furthermore, the analysis provided an understanding of risk to the special cultural fabric of the monument, both from the DDT and from remedial measures. This assessment document served as a basis for the regulators, trustees, and the park service to develop an approach to discuss with other interested parties. The result was a successful resolution of the issue with minimal remediation, no harm to the cultural resources, and support of all parties.

3. Risk Significance

Frank Young (1987), the former head of the FDA, sees these issues as follows:

"No activity is completely safe, no action taken is without risk."

As he describes it, because absolute safety makes no sense, resources need to be focused on significant risks. What is significant? From the general legal and regulatory viewpoint, it is necessary to set a reasonable safety standard equivalent to a level posing virtually no harm (or human health or environmental risk). However, what should that level be, based on what known data, and containing how much uncertainty and indeterminacy? This is Vincent's (1999) point as mentioned earlier: the decision process of environmental management is not an objective, scientific process, but a value-laden, political process, which affects real "people at risk." Corporate managers must understand and effectively engage people's concern about risks that they cannot control themselves. In reality, the corporation and regulators are in "direct control," that is, until stakeholders jeopardize their credibility and force "redirection" as seen at sites like the Pine Street Barge Canal (Strasser, 2000).

Using the definition of risk as harm from the foregoing evaluation, the corporate manager needs to develop an understanding of the significance of the risk as it is placed into a local context of space and time. This context helps those involved understand the relation of the defined risk system to the overall population of people, the range of people affected, the environment, and critical biotic, abiotic, and cultural components (review Figures 1.1 and 1.2). Synoptically defining risk significance requires an evaluation of multiple scales of risk as harm (USEPA, 1991a and b, Harris and Harper, 2001, Canter, 1996, and Jain *et al.*, 1993), and what follows is one structural approach to achieve this goal. The manager can characterize the significance of risk as "harm" using several categories of "health:"

- **Public Health—known or suspected carcinogens** (following 40 CFR 300.430 [E][2][i][A][2]):
  - Not Significant to Trivial = incremental lifetime cancer risks below 1×10^{-6} (0.000001 or one-in-one-million persons so exposed) for both individual chemical exposures and cumulative exposure scenarios.
  - Evaluative Significance = incremental lifetime cancer risks between 1×10^{-6} (0.000001 or one-in-one-million persons so exposed) and 1×10^{-4} (0.0001 or one-in-ten thousand persons so exposed) for both individual chemical exposures and cumulative exposure scenarios.
  - Defined Significance = incremental lifetime cancer risks above 1×10^{-4} (0.0001 or one-in-ten thousand persons so exposed) for both individual chemical exposures and cumulative exposure scenarios.

- **Public Health—noncarcinogens** (or those chemicals with a systemic effect) (following 40 CFR 300.430[E][2][i][A][1]):
  - Not Significant to Trivial = exposures less than the reference dose— *i.e.*, exposures, for both individual chemical exposures and cumulative exposure scenarios, with a noncancer Hazard Quotient less than unity (one or 1.0)—will generally not be associated with health risks. (Where the reference dose is defined as "an estimate (with uncertainty spanning perhaps an order of magnitude) of a daily oral exposure to the human population (including sensitive subgroups) that is likely to be without an appreciable risk of deleterious effects during a lifetime" [USEPA, 1999c].)
  - Evaluative Significance = exposures exceeding the reference dose— *i.e.*, exposures, for both individual chemical exposures and cumulative exposure scenarios, with a noncancer hazard value greater than unity (one or 1.0)—may be associated with adverse health effects in a population. Nonetheless, a clear distinction that would categorize all exposures below the reference dose as acceptable (*i.e.*, risk-free) and all exposures above the reference dose as unacceptable (causing adverse effects) cannot be made.

- **Ecotoxicological Health**—the following definition of significance in terms of ecological impact from chemicals to specific ecological receptors is based upon the general U.S. environmental policy (see 40 CFR 121(b)(1) and (d) and USEPA, 1997e):

- ○ Not Significant = exposures less than the ecotoxicological benchmark—*i.e.*, exposures, for both individual chemical exposures and cumulative exposure scenarios, with an Ecological Hazard Quotient less than unity (one or 1.0)—will generally not be associated with environmental risks.
- ○ Evaluative Significance = exposures greater than the ecotoxicological benchmark—exposures, for both individual chemical exposures and cumulative exposure scenarios, with an Ecological Hazard Quotient greater than unity (one or 1.0)—will generally be associated with a potential risk for adverse effects from exposure to stressor (chemical, physical, biological, or others). The magnitude of the Ecological Hazard Quotient is generally accepted as indicating a relative risk to the end-point species (for example, wildlife, plant, or fish) under evaluation. By referring to the percentages of exposure resulting from different pathways (*e.g.*, food ingestion, soil ingestion, water ingestion), the relative contribution to total potential risk for each exposure pathway can be identified. Confidence in the Hazard Quotient increases with greater certainty in the organism's exposure concentration and the toxicity reference value or benchmark. The greater the confidence one has in the predictive value of the Hazard Quotient, then the more certain the pass/fail decision point.

- **Socio-Cultural and Socioeconomic Health**—for the particular system, element, process, or attribute of concern (the measures are extended based on Jain *et al.*, 1993 and Canter, 1996):
  - ○ No harm or benefit:
    - Nature (impacts/benefits have low probability of occurrence, few if any are affected, covers only a small area and/or occurs over short duration).
    - Severity (is low in terms of local sensitivity to it and it has a low magnitude of impact).
    - Potential for Mitigation (impact is reversible and significant institutional capacity exists for dealing with it).
  - ○ No Adverse Impact—impact does not diminish the functional quality or integrity of appropriate characteristics for the system, element, process, or attribute of concern:
    - Nature (impact has very low probability of occurrence, few if any affected, and very small area and/or occurs over short duration).
    - Severity (is low in terms of local sensitivity to the impact and it has essentially a zero magnitude of harm or impact).
    - Potential for Mitigation (impact, if any, is immediately reversible and institutional capacity exists for dealing with it).
  - ○ Defined Significance—a possible effect that could diminish the functional quality or integrity of appropriate characteristics, again for the item or issue of concern:

- Nature (the impact has a defined probability of occurrence, more than a few are affected, and it covers a sizeable area and/or occurs over an extended duration).
- Severity (is moderate to high in terms of local sensitivity and it has at least a moderate magnitude of harm or impact).
- Potential for Mitigation (impact may not be reversible and the institutional capacity may be insufficient for dealing with it).

We have attempted to define a measure of significance here using terms and occasional numeric standards common in U.S. environmental policy. However, what do we seek to achieve by such delimiting terms? Harris and Harper (2001) suggest that commonplace terms be used, such as perturbation, harm, injury, severe/irreparable injury, and catastrophic. Their purpose in choosing such terms is to anchor any discussion among experts and stakeholders for consensus. Our point here is to provide a framework that can serve the corporate manager in understanding how others within and without their organization will view the significance of the environmental peril they confront. Developing an appreciation of this aspect of environmental risk management is vital to successfully defining risk and ultimately managing it.

4. Appreciating Uncertainty

Facts, information, and the usual testing and analysis have variability, uncertainty, and indeterminacy in each of them, and this is particularly true with risk assessments. Unfortunately, people often believe (even before coming to hear about your property) that the environment is polluted and degraded. Therefore, many people are predisposed to equate even more readily the norms of uncertainty and indeterminacy in science as an "inability...or unwillingness to manage the environment" (Llewellyn, 1998). Such a situation corrodes trust. Thus, according to Carpenter (1995), it is necessary to explain, in lay terms:

- What we know and the degree of confidence of that knowledge, and what is directly measured, estimated, and/or conjectured, and on what basis were the judgments (scientific, professional, or guessing) founded.

- What we do not know or are unsure of and why.

- What else could we know? This question answers the value of information and the price one must pay to obtain or create it.

- What is it that we should know to act with these uncertainties?

Therefore, the corporate manager must take time to build a common sense, logical argument about the risk system for stakeholder consideration, debate, and negotiation. Management should incorporate thoughtful and appropriate comparisons, seeking to educate and willing to be educated about the situation so that participative decision-making proceeds reasonably and rationally.

C. MANAGING THE DEFINITION OF RISK

We began this chapter defining risk management following Rejda (1992); and we have seen that loss exposures from environmental impairment are a complex type of liability, primarily because of the problem in the definition of the risk to be managed. Environmental decisions must consider a litany of perspectives and demands from within (Figure 2.8) and without the organization. Should the corporate manager allow others to define the risk? Regulators think so, and the law supports them. Stakeholders and especially the neighbors of the impaired property think they should. Does the existing risk assessment/management/communication paradigm suffice? Will we be saved from unworthy and costly cleanups by RBCA? If the corporate managers cannot make the decision themselves, how can they successfully engage others in this most complicated process?

Portney (1991) identifies two principal styles of public environmental decision-making, namely:

1. "Positivist"—an individualistic approach founded upon the utility of analytical methods with two important types:

    o cause and consequence, which employs quantitative tools to understand historical causes of environmental problems and/or consequences of environmental decisions; and

    o prescriptive, an analytic approach using techniques such as probabilistic risk assessment to objectively determine future outcomes.

2. Public Policy-Making—This is the classic process: understanding the problem → developing solutions (formulation and adoption) → putting solutions into effect (implementation) → testing success and making it stick (evaluation and, if necessary, reformulation to start-off all over again). It involves a large cast of players, presumes that any number of alternatives is possible, and embraces the impact of politics through interpersonal and group interactions based on their values.

Most managers are familiar with the positivist approach, and the prescriptive type in particular, and depending upon our training and experience, we likely favor it. Certainly, Charnley's (2000) analysis suggests business and the technical consulting/engineering realm favor it. However, the issue here is not whether one approach is better than another. Rather, it is ensuring a process that helps the corporate manager perceive the critical issues early enough (through an activity called "front-loading") and with sufficient detail to be better informed, focused, and ready regardless of the decision-making process. Front-loading in this context means early, in-depth consideration of the many scenarios that might evolve over the course of a matter (Tissembaum, 1993). It requires "preliminary" studies before the full, formal processes begin.

The currently used methods of incorporating expert and lay knowledge, information, perceptions, and values into environmental decisions include (Portney, 1991):

- public policy model (previously mentioned),

- risk communication (discussed in more detail in Chapter 3 and Appendices A and C), and

- dispute resolution (negotiation and mediation, such as the resolution of the Pine Street Barge Cannel Superfund site [Strasser, 2000]).

An interesting recent development that attempts to use all three of these methods to achieve legitimate and efficient environmental health risk management decisions is the framework for environmental health risk management developed by the Presidential/ Congressional Commission on Risk Assessment and Risk Management (or PCCRARM, 1997a, 1997b, and 1997c). A clear need to modify the traditional approaches used to assess and reduce risks emerged as a major theme from the commission's deliberations. According to the commission, traditional approaches rely on a chemical-by-chemical, medium-by-medium, risk-by-risk strategy. In so doing, they focus attention on refining assumption-laden mathematical estimates of the small risks associated with exposures to individual chemicals, rather than on the overall goal of reducing risk and improving health status. For this reason, the commission sought to create a framework to guide investments of valuable public sector and private sector resources in researching, assessing, characterizing, and reducing risk. The framework integrated the extant risk paradigm (discussed earlier in this chapter and in Appendix A) with a public policy-formation model. It sought a process that would balance "good science" with sound policy formation, while actively engaging stakeholders. The resulting framework is a classic six-stage decision or problem-solving process for risk management (Figure 2.9). The expectation is that the framework was scalable to the importance of a public health or environmental problem, and which would help all types of risk managers— government officials, private sector businesses, individual members of the public. The Commission's framework is useful in that it:

- applies an *integrated* team approach working toward a consensual solution,

- seeks *interaction* opportunities among interested parties and the multi-disciplinary team, and

- allows for subsequent process *iterations*, if necessary.

We believe that process is substance and that the process *is* the risk management decision! Regardless of the paradigm, framework, or regulatory decision construct (modeled after RBCA or not), we believe that it is essential that the corporate manager be prepared to work within it so that a well-implemented process yields an optimal decision product satisfying:

- the company, by reducing its financial and legal risks at the lowest possible investment of time and money;

- the public, by protecting them and improving the community in which they live;

- the regulators, by protecting human health and the environment; and

- other stakeholders, by meeting mutually accepted goals.

From the foregoing discussion, we conclude that while corporate managers should not abandon either good science (Figure 2.1) or sound business calculus (suggested by several figures in this chapter and summarized in Figure 2.8), they must strengthen their environmental risk management system by managing the definition of environmental risk in the following ways (Figure 2.10):

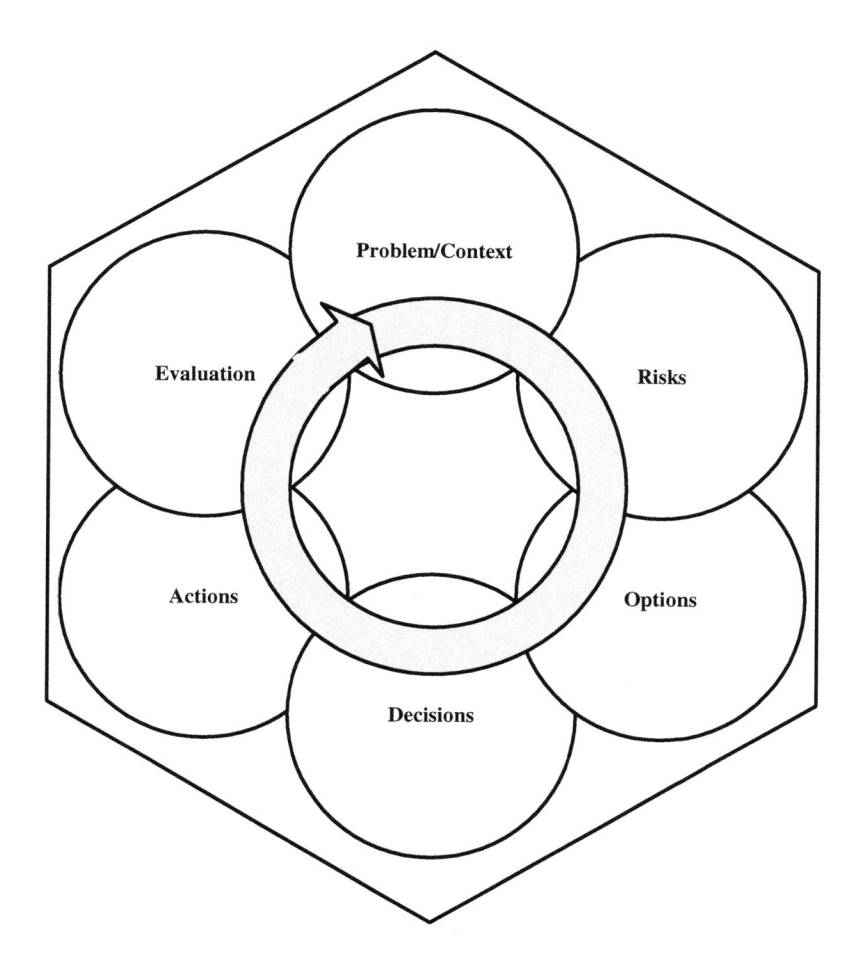

Figure 2.9. Framework for Risk Management as proposed by the Presidential/Congressional Commission on Risk Assessment and Risk Management (after page 3, PCCRARM 1997a).

- identify and define the peril as a loss exposure (Figure 2.1) from a multidimensional view in terms of hazard, risk, and cost, and one that measures risk (both actual and perceived) synoptically (as harm, with importance and significance, and containing uncertainty);

- understand and network together the internal bases of knowledge or "players" surrounding or involved in the matter within the corporation and successfully engage them to inform the environmental risk management process (Figures 2.6 and 2.8);

- apply risk assessment methods that integrate information from all aspects of the company (Figures 2.2 through 2.5) and combine it with the scientific and technical assessment (Figures 2.1);

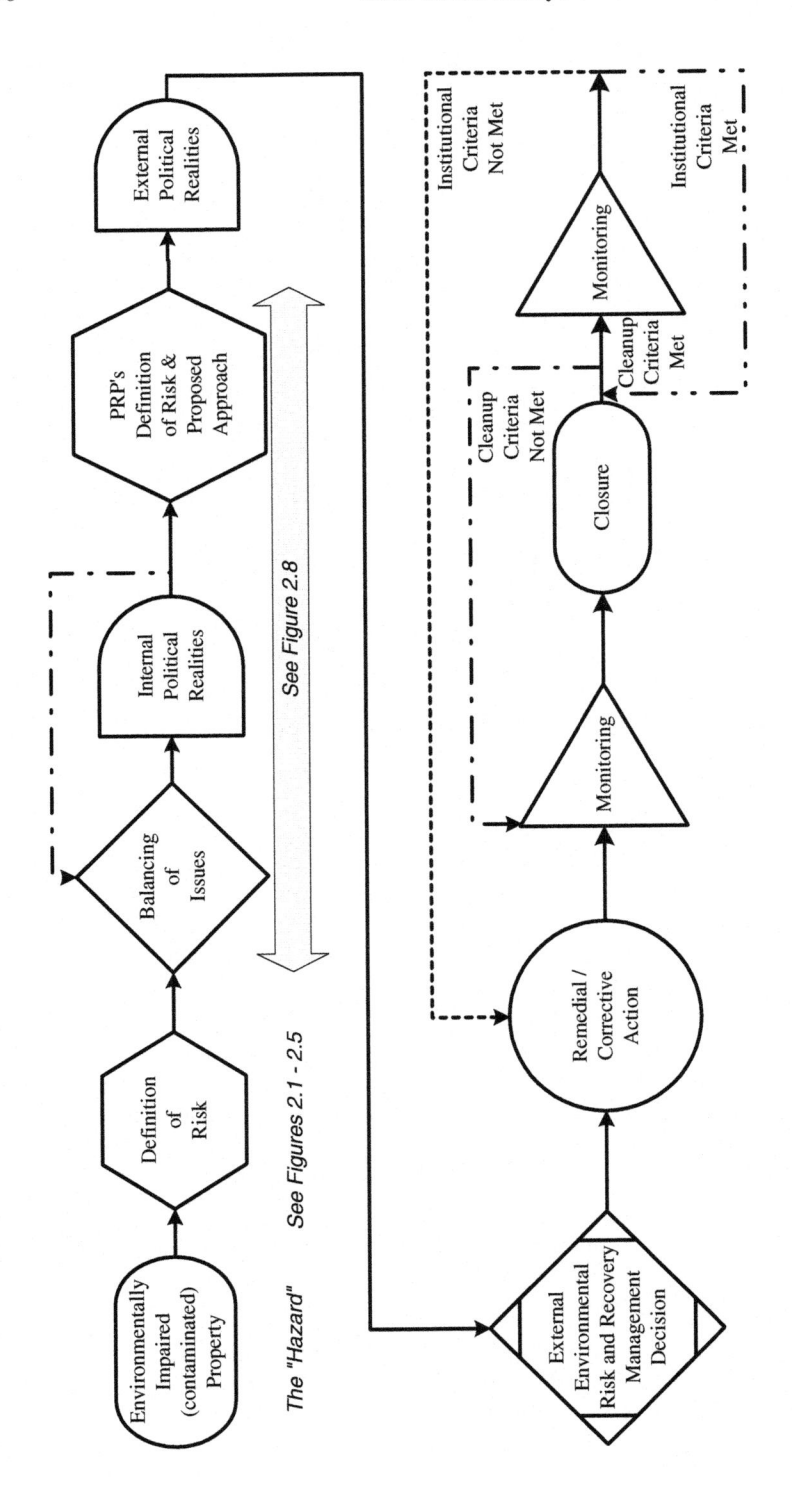

Figure 2.10 Overview of the Environmental Risk Management Process.

- develop a sound set of risk management options based on this information and with an understanding of the various "players" surrounding or involved in the matter beyond or outside the corporation (Figures 2.6 and 2.8); and

- be prepared to engage the external parties using a communications approach that argues from a position of knowledge-strength, based on a synoptic understanding of the definition of risk.

Most importantly, it is crucial to do, or at least start doing, these things as early as possible! This is because once the remedial or corrective action process "hits the streets" the definition of risk becomes much more difficult to manage.

Subsequent chapters of this book outline a natural evolutionary step in the environmental risk management paradigm. Instead of looking at the issue of environmental risk as a regulatory compliance protocol or merely achieving a strict numerical goal, we suggest designing environmental risk to redevelop or restore impaired and stigmatized properties. This transcends the typical "cookie-cutter" steps when reckoning critical factors for success because it starts early with a positive end in mind. The many technical, regulatory, legal, property, financial, company image/relationship complexities require *integration* to achieve results. Further, the dynamics of the situation warrant capability beyond traditional practices. For example, although each site has its own unique attributes, identifying site barriers to revitalization potential is an initial, classic step that needs to be completed in virtually all cases of impaired property. These barriers can become components in an economic and environmental model to prepare for site revitalization. The following items are some common barriers:

- divergent stakeholder objectives,

- lack of predictable cleanup funding or cost-recovery mechanisms,

- lack of uniformity in environmental/economic perspectives,

- risk-averse influences, and

- a bureaucratic versus performance-driven process.

The evolutionary step suggested here we believe will help the corporate manager construct an approach and use the knowledge-based network needed to reconcile risk-based and precautionary viewpoints, identify common barriers, seek economic catalysts with potential to aid risk mitigation, and help the corporate manager in managing the definition of risk. This evolutionary step is Risk-Based Analysis.

Chapter 3

INTERFACING THE ASSESSMENT, MANAGEMENT, AND COMMUNICATION OF RISK

Cris Williams, Kurt A. Frantzen, and Judy Vangalio

I. INTRODUCTION

We have considered the integrated and complicated nature of environmental risk management. In doing so, we identified a critical management issue: the various definitions of environmental risk that arise within and without the corporation. Additionally, we saw what Toll (1999) describes as the inseparability of the scientific and socioeconomic dimensions of environmental problems. To deal with this reality, we identified the need for a front-loaded management process (Figure 3.1) using good science, objective/subjective integration, and sound business calculus that is able to:

- identify and define the peril as a loss exposure from a holistic, multidimensional perspective;

- network with various sources of knowledge and "players" within and without the company to understand their goals and values in order to inform the process;

- apply integrative assessment methods to define and understand the risk system within its operational abiotic, biotic, and cultural environment;

- develop management options to control the risk system based upon this information, allowing the manager to seek mutual gains with involved stakeholders; and

- apply a communications approach, based upon a synoptic understanding of the definition of risk, that argues from a position of knowledge-strength and follows the process principles of engagement, explanation, and expectation clarity (Chan Kim and Mauborgne, 1997).

The evolution of the assessment/management/communication paradigm is the result of decades of deliberation by experts in government, academia, and the private business sector. Despite their laudable efforts, environmental managers still must grapple with the bipolar realities of risk through their own management systems to avoid ineffective decision-making resulting in unending cost-escalation, infinite liability, a plummeting corporate image, and loss of regulatory compliance. While often seen and treated as separate aspects of a larger process, the manager must be able to effectively interface with the paradigmatic troika of risk assessment–risk management–risk communication, with the help of an external (to the corporation) and substantial knowledge-based network of consultants, regulators, lawyers, politicians, and the media, as well as interested people.

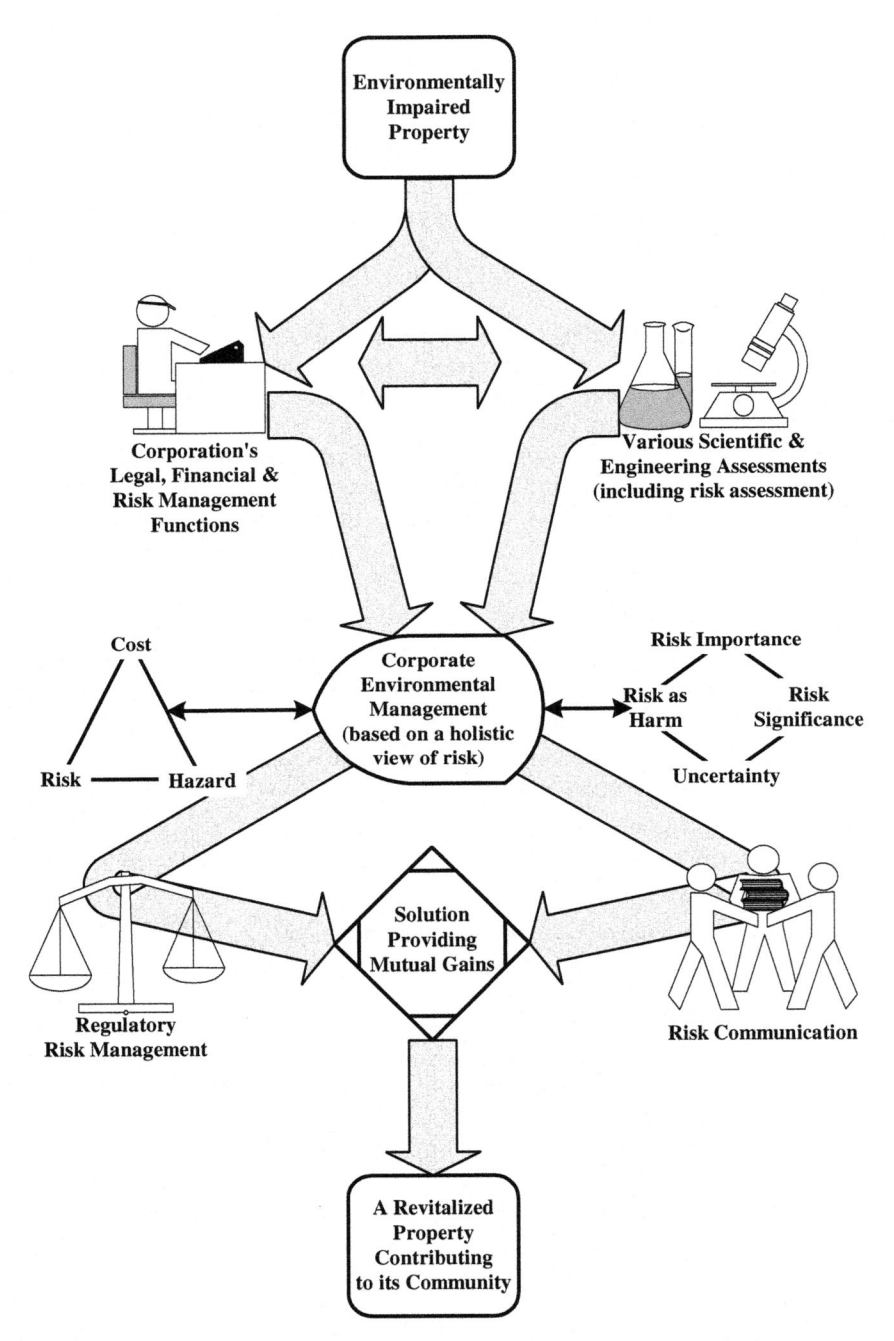

Figure 3.1 Integrated environmental risk management process incorporates a holistic view.

In the previous chapter, we defined environmental risk in business, health, and ecological terms. We further outlined several important concepts of environmental risk management associated with impaired property. In this chapter, we build on this foundation by discussing protocols and policies behind the actual application of the risk assessment–risk management–risk communication paradigm to demonstrate a management approach for interfacing these distinct operations, an approach that we believe is best served through Risk-Based Analysis.

II.   RISK ASSESSMENT: THE JOB EVERYONE HATES

Risk assessment, at least to us, seems to be a job that everyone hates. This dislike stems from many factors. Most risk assessors have failed to lessen the loathing by not instilling "transparency" and "clarity" to the process as directed by USEPA's 1995 Policy for Risk Characterization (USEPA, 1995a). Risk assessment is often a scientifically and mathematically intensive exercise, with myriad steps and opportunities for confusion on the part of any audience. Hate or rejection is a natural reaction to a complicated process, especially if that process and its results are poorly articulated. Regardless, environmental protection as a matter of problem-solving begins with risk assessment as an agent of information development. The establishment of a solid basis of information is the first step in the goal of properly framing environmental liability for the corporate manager.

In seeking to better interface the components of the risk paradigm, the following discussion (along with Appendix A) aims to outline the process of risk assessment and the protocols followed in assessing risks to human health and ecological resources. This section of Chapter 3 concludes by reviewing the different kinds of information developed through risk assessment and the products it provides. Finally, the discussion of risk assessment is not exhaustive but highlights what the authors believe to be important. We acknowledge that others may disagree with our selection or presentation of the material, and for that reason we urge the reader, who may need additional background, to use the many resources and references cited herein, as well as the many excellent texts available on the subject.

A. HUMAN HEALTH RISK ASSESSMENT

Human health risk assessment is a formal protocol for qualifying and quantifying potential threats to human health as they may emerge from exposure to chemical, biological, or physical agents in any environmental media (NRC, 1983). As described in Appendix A, the National Academy of Sciences (NRC, 1983) specifies four steps in the risk assessment process (Figure 3.2): hazard identification, exposure assessment, toxicity assessment, and risk characterization

As currently practiced, risk assessment is an amalgam of the discipline of toxicology ("the basic science of poisons," Amdur *et al.*, 1991), the practice of assessing exposure, and numerous policy decisions permeating each step in the protocol. Scientists (often toxicologists) may view risk assessment as good science corrupted by policy. The policy maker, regulator, and the public may be of the opinion that all the science in risk assessment is unnecessarily confusing and is

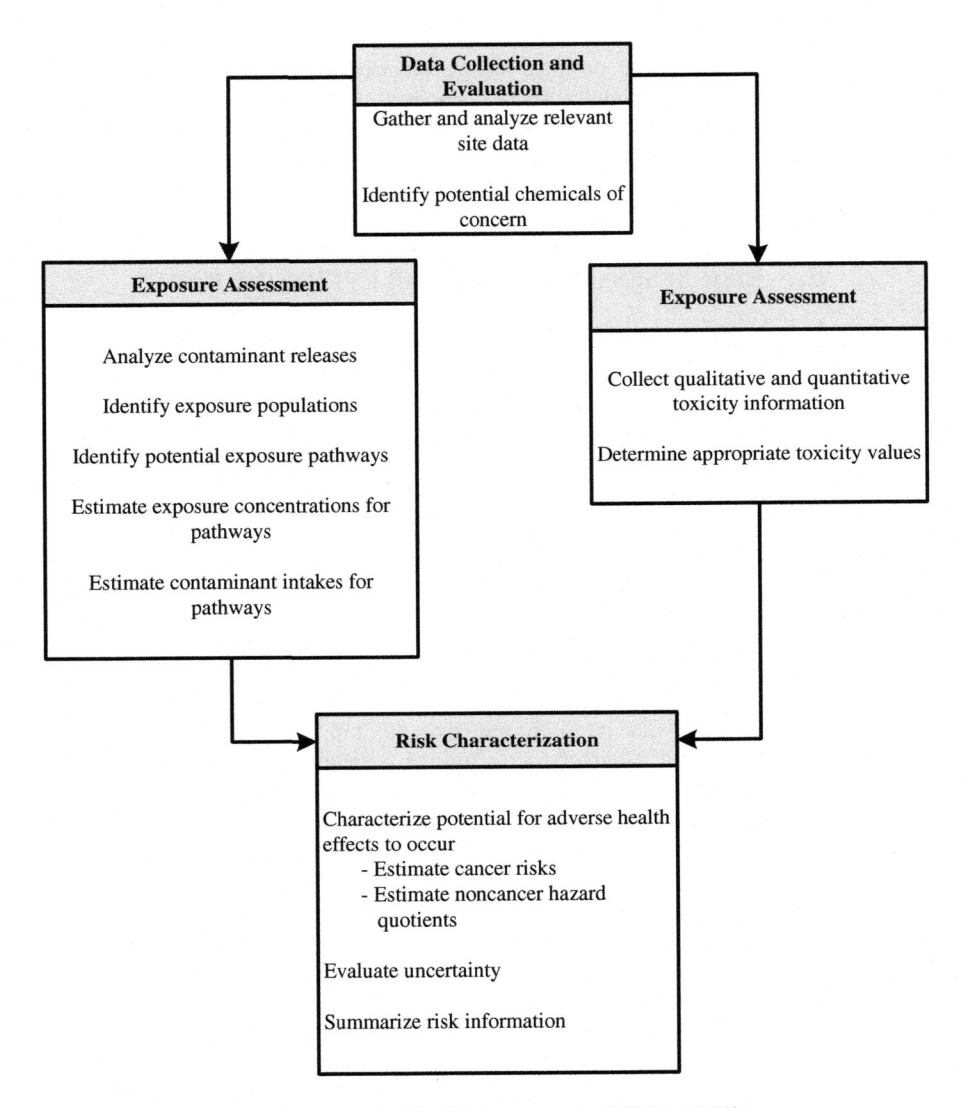

Figure 3.2 Overview of the Human Health Risk Assessment of USEPA (1989b).

simply used to justify less rigorous cleanup. To help the corporate manager develop a better process, we briefly discuss some of the major science, exposure, and policy elements critical to an appreciation of the role risk assessment plays in the process.

1. The View from Toxicology

From a toxicologist's point of view, risk assessment is full of policy decisions with little basis in science. Some decisions reflect the way threats to human health posed by carcinogenic (cancer-causing) substances are evaluated. Other policy decisions impact how noncarcinogenic (systemically acting) substances are assessed.

a. Carcinogenic Substances

For ethical and financial reasons, much of what we know about the ability of chemicals and other substances to cause cancer in humans comes from laboratory studies of rats and mice. To observe sufficient numbers of tumors in animals treated with a suspected carcinogen, the study uses many animals and high doses of the substance. Animals treated with high doses raises two critical issues regarding the human applicability of such studies:

- The *qualitative* relationship of the response in animals to humans; is it the same? The following discussion exemplifies these issues.
    - ○ It should be intuitive that humans are not simply "big rats" and that carcinogenic responses in animals are not necessarily predictive of carcinogenic responses in humans. Although animals are often reliable *qualitative* predictors of cancer in humans, numerous substances cause tumors in treated animals but not in exposed humans. For example, the chlorinated hydrocarbon methylene chloride has been demonstrated to cause liver tumors in rats and mice but, according to the USEPA, evidence for carcinogenicity in humans is "inadequate" based on studies of persons exposed to methylene chloride in the workplace (USEPA, 2001b). This finding relates in part to the fact that differences exist between the way in which methylene chloride is metabolized in rats and mice compared to humans (ATSDR, 1999).
    - ○ The use of susceptible animals in cancer bioassay studies also responds to qualitative differences in carcinogenic influences between animals and humans. For example, many chemicals that cause liver tumors in animals do so by virtue of the fact that the animals are selectively bred to be susceptible to liver cancer or, because of repeated inbreeding, inadvertently possess a high "background" incidence of liver tumors. This factor increases the chance of observing liver tumors in animals treated with suspected liver carcinogens and thus allows the use of a smaller number of animals in the study, making the study less expensive to conduct. However, the relevance of cancer in these animals to cancer in humans becomes suspect. The chlorinated hydrocarbons trichloroethylene and tetrachloroethylene are considered "probably carcinogenic to humans" by the International Agency for Research on Cancer (IARC) on the basis of studies with such susceptible animals (IARC, 2000).
    - ○ Certain substances have been determined to be carcinogenic in humans even though there are fundamental anatomic differences between humans and the animals species chosen to be representative of the human carcinogenic response. For example, polycyclic aromatic hydrocarbons (PAHs), a group of chemicals found at many hazardous waste sites (and in many medicated shampoos), cause tumors in the forestomach in rats and mice (USEPA, 2000). Consequently, PAHs are classified by the USEPA as "probable human carcinogens" even though humans do not have a forestomach.

- The *quantitative* relationship between the dose used to treat animals and the dose humans might receive if exposed at levels found in the environment; again, is it the same?

  o The Environmental Protection Agency (USEPA) and most state regulatory agencies use a model known as the linearized multistage (LMS) model to extrapolate from the high (maximally tolerated) dose an animal receives in a lifetime carcinogenicity study to lower doses that do not cause cancer in animals but may be more like the low-level exposures humans typically experience. This model predicts the "slope" of the experimentally derived dose/carcinogenic response curve and provides an estimate of carcinogenic potency. Chemicals with steep dose/response curve slopes are more potent carcinogens than those with shallower slopes. The Agency has concluded on theoretical grounds that cancer follows a series of discrete stages (initiation, promotion, and progression) that ultimately result in uncontrolled cell proliferation (cancer). Consistent with this conclusion, the use of the LMS model permits an estimation of a slope or potency factor that is not likely to be exceeded if the real slope could be measured. Compelling scientific arguments can be made, however, for several other extrapolation models that, if used, could result in significantly reduced values for the slope factor, many times lower than those estimated using the LMS model. This premise has *quantitative* implications concerning how carcinogenic potency is estimated in human health risk assessments. Existing USEPA slope factors calculated using the LMS model represent upper-bound values based on animal data, which may not necessarily predict actual human cancer potencies.

b. Noncarcinogenic Substances

Like cancer, much of our knowledge regarding the chemicals causing toxic effects other than cancer (such as liver or kidney toxicity, birth defects, reproductive effects, and neurotoxicity) comes from animal studies. This reliance requires the use of simplifying assumptions, *e.g.,* effects in animals are similar in humans, the effects can be extrapolated between routes of exposure, and the same effects can be extrapolated over varying periods of exposure. As with cancer-causing substances, noncarcinogenic effects observed in an animal species or by one route of exposure may not occur in humans or by another route, or they may occur at a higher or lower dose due to differences in the pharmacokinetics (*i.e.,* the absorption, distribution, metabolism, and excretion) of a compound between species or when exposure occurs by different routes.

Regulatory policy dictates that uncertainty associated with such assumptions is accounted for using "safety" and "modifying" factors.

- "Safety factors" reflect the uncertainty associated with species-to-species extrapolation, the need to assure the safety of sensitive individuals, and the necessity to account for animal study duration of insufficient length to reflect likely human exposures.

- "Modifying factors" are used to include a quantitative professional assessment of additional uncertainty associated with the selection of the critical animal study, as well as uncertainty in the entire database for a given chemical substance not explicitly addressed by other factors.

These factors are conservative (or health-protective) as they generally overestimate the uncertainties associated with the assessment of toxic effects (USEPA, 1989b). For example, estimates of safe doses for the chemicals dimethylaniline and acrolein are based on studies in which the toxicological endpoints (blood effects and irritation for dimethylaniline and acrolein, respectively) were determined in studies in animals (USEPA, 2000). Large (1,000 or greater) safety factors were used to estimate safe doses for these chemicals in part because available animal studies were of a short-term (rather than chronic) nature (13 weeks for dimethylaniline and 62 days for acrolein) (USEPA, 2000). However, long-term human occupational studies exist for both chemicals, including studies that consider the same critical effects as USEPA's preferred animal studies (Williams *et al.*, 1995). Use of available human data for certain chemicals eliminates the need for large safety factors and reduces the uncertainty associated with estimating the safe dose.

2. Defining Exposure

The key to any risk assessment lies in defining the exposure (Figure 3.3):

- who is/was/will be exposed,
- how they are/were/will be exposed,
- to what substances they are/were/will be exposed,
- for how long they were/will be exposed, and
- how intensively they were/will be exposed.

Problem Formulation is a scoping effort that helps the risk assessor define these parameters and focus the risk assessment. It results in the following:

1. Assessment Endpoints, which are based on potentially complete exposure pathways and toxicological effects, and which require two elements for their definition:
   - the identification of the particular human population of interest – residential or worker populations, for example.
   - the characteristics about what is potentially at risk and important to protect, *i.e.*, human health.

   With the Assessment Endpoints in hand, one must measure whether or not a potential threat exists. The Measurement Endpoint is the potential level of exposure, presence of symptoms, or hospital admissions indicating a measurable threat.

2. A set of "Conceptual Models" are then prepared that describe the system under assessment. Although many different models are useful, two models of greatest value are briefly defined here and more fully explained in Chapter 4:

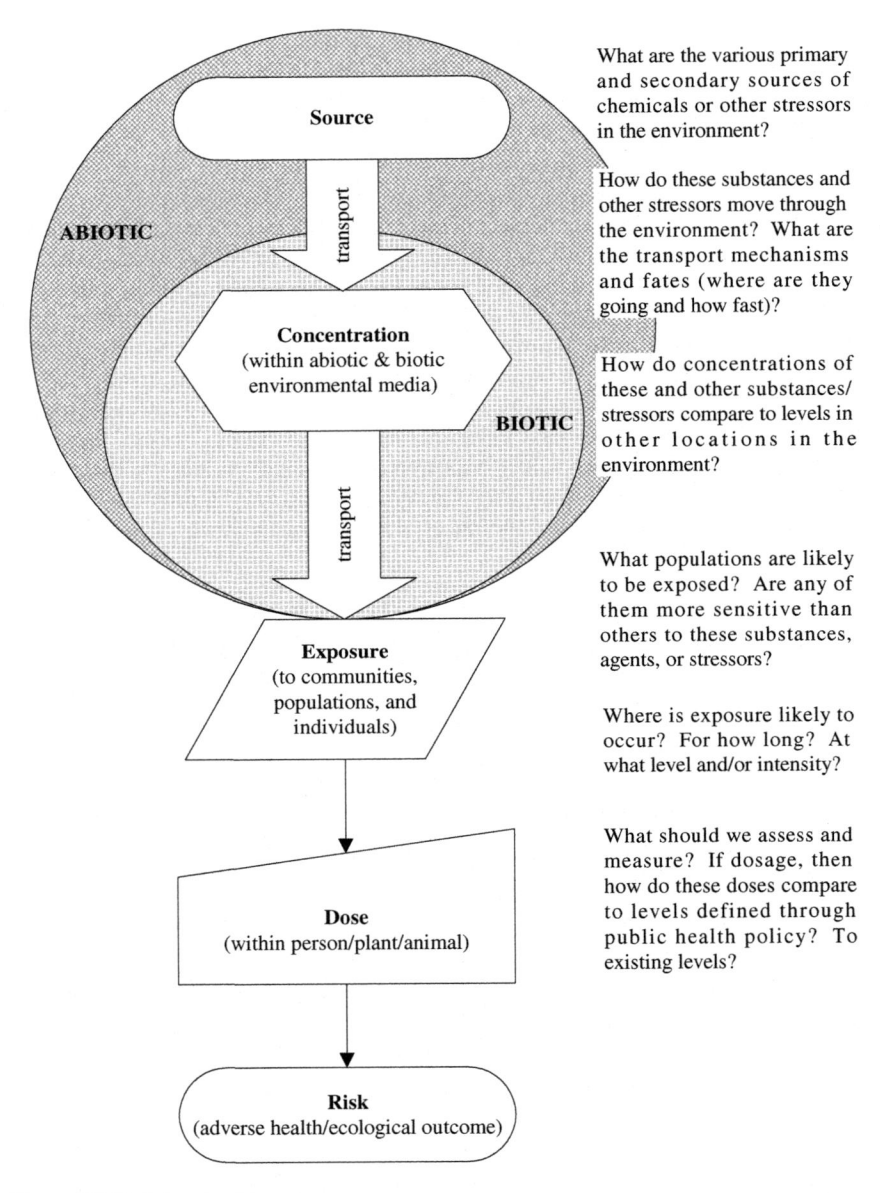

What are the various primary and secondary sources of chemicals or other stressors in the environment?

How do these substances and other stressors move through the environment? What are the transport mechanisms and fates (where are they going and how fast)?

How do concentrations of these and other substances/ stressors compare to levels in other locations in the environment?

What populations are likely to be exposed? Are any of them more sensitive than others to these substances, agents, or stressors?

Where is exposure likely to occur? For how long? At what level and/or intensity?

What should we assess and measure? If dosage, then how do these doses compare to levels defined through public health policy? To existing levels?

Figure 3.3 Defining the risk system.

- o   Conceptual Site Model (CSM)—this model graphically describes a site and its physical environs, with the intent to identify potential sources of chemical or other substances and provide insight to the potential transport and fate of released chemicals into the environment.
- o   Conceptual Risk System Model (CRSM)—this model provides a graphical depiction of the spatial and temporal relationships (now and

in the future) among sources, the environmental fate of substances, and the potentially exposed human population(s).

Together, the CSM and CRSM capture, in a graphic format, the critical elements of exposure that help to define the risk assessment.

3. Characterizing Risk

As alluded to previously, chemical risks are generally assessed in one of two ways: as potential carcinogens and/or as noncarcinogens. The assessment follows two separate but complementary tracks because:

- noncarcinogens generally exhibit a threshold dose below which no adverse effects occur,

- whereas policy allows no such threshold to be used for potential carcinogens (USEPA, 1986a, 1989b, and 1995b).

As used here, the term carcinogen means any agent for which the USEPA has determined there is sufficient evidence that exposure may result in continuing, uncontrolled cell division (cancer) in humans and/or animals. Conversely, the term noncarcinogen means any agent for which the carcinogenic evidence is negative or insufficient. Exposure to some agents may result in both carcinogenic and noncarcinogenic effects. In these instances, both the carcinogenic and noncarcinogenic effects are evaluated.

a. Individual Risks for Carcinogens

Scientists generally have been unable to demonstrate experimentally a threshold for carcinogenic effects. Federal regulatory agencies assume that any exposure to a carcinogen theoretically entails some finite risk of cancer (USEPA, 1989b). However, depending on the potency of a specific carcinogen and the level of exposure, such a risk could be vanishingly small.

In evaluating agents for carcinogenicity, USEPA, for example, uses a two-part assessment involving a weight-of-evidence classification and a quantitative determination of carcinogenic potency (cancer slope [potency] factors or CSF). The weight-of-evidence classification reflects available data, adequacy of studies, types of studies, and observed responses.

CSFs are used to predict the potential number of excess cancers that will arise in response to lifetime exposure to an agent. For reasons previously discussed, CSFs are predominantly based on animal bioassay data, although human epidemiological data (*i.e.*, data gathered from studies of defined human populations under somewhat controlled environments, such as factory workers exposed to benzene or vinyl chloride) are preferred and are used when available.

Using CSFs, lifetime excess cancer risks for individual chemicals can be estimated by the following mathematical expression:

$$Cancer\ Risk = \sum Intake \times CSF$$

Where:

- Intake = chemical-specific ingestion, dermal, or inhalation intake

- CSF = chemical- and route-specific cancer slope (potency) factor

Carcinogenic risks for the ingestion (oral), dermal, and inhalation routes of exposure are calculated and added (if applicable) as follows:

$$Total\ Excess\ Cancer\ Risk = Intake_o \times CSF_o + Intake_d \times CSF_d + Intake_i \times CSF_i$$

Where:

- the "o" subscript denotes the oral route of exposure;
- the "d" subscript indicates the dermal route of exposure; and
- the "i" subscript denotes the inhalation route of exposure.

USEPA has adopted a policy (that most state regulatory agencies follow) for making decisions about what is and what is not an acceptable exposure to known or suspected carcinogens (40 CFR 300.430[E][2][i][A][2], USEPA 2001a). Risks ranging from 1×10^{-4} (0.0001 or one in 10,000 persons) to 1×10^{-6} (0.000001 or one in one million persons) are generally considered acceptable by the USEPA.

b. Individual Risks for Noncarcinogens

Noncarcinogenic effects are assessed by comparing the estimated average exposure to the acceptable daily dose, referred to by the USEPA as the reference dose (RfD) (USEPA, 1989b). The RfD is a provisional estimate (with about an order of magnitude of uncertainty) of a daily exposure to a human population, including sensitive subgroups, which is likely to be without an appreciable risk of deleterious effects during a lifetime (USEPA, 2001b). USEPA selects the RfD by identifying the highest reliable No Observed Adverse Effect Level (NOAEL) or Lowest Observed Adverse Effect Levels (LOAEL) in scientific literature. They then apply a suitable uncertainty factor (described above and typically ranging from three to 10,000) to allow for differences between the study conditions and the human exposure condition to which the RfD is to be applied. NOAELs and LOAELs can be derived from either human epidemiological studies or animal studies; however, they are usually based on laboratory experiments in animals in which relatively large doses are used. Consequently, uncertainty factors are applied when deriving RfDs to compensate for data limitations inherent in the underlying experiments and for a lack of certainty created by extrapolating from high doses in animals to lower doses in humans.

Hazards associated with noncarcinogenic effects are assessed by comparing the estimated exposure to the acceptable daily dose, referred to as the RfD by USEPA. Noncarcinogenic hazards are assessed by calculating a hazard quotient, which is the ratio of the estimated exposure to the RfD as follows:

$$Noncancer\ Hazard = \sum \frac{Intake}{RfD}$$

Where:

- Intake = chemical-specific ingestion, dermal, or inhalation intake;
- RfD = chemical- and route-specific reference dose

Noncancer hazards for the ingestion (oral) and dermal routes of exposure are calculated and added (if applicable) as follows:

$$Total\ Noncancer\ Hazard = \frac{Intake_o}{RfD_o} + \frac{Intake_d}{RfD_d} + \frac{Intake_i}{RfD_i}$$

Where:

- The "o" subscript denotes the oral route of exposure;

- The "d" subscript indicates the dermal route of exposure; and

- The "i" subscript denotes the inhalation route of exposure.

As for carcinogens, USEPA has adopted a policy for noncarcinogens for making decisions about what is and what is not an acceptable exposure (40 CFR 300.430[E][2][i][A][1]; USEPA, 2001a). Exposures that are less than the RfD—*i.e.*, exposures with a noncancer hazard value less than one—will not be associated with health risks. Exposures exceeding the RfD, *i.e.*, exposures with a noncancer hazard value greater than one, may be associated with adverse health effects in a population. Nonetheless, a clear distinction that would categorize all exposures below the RfD as acceptable (*i.e.*, risk-free) and all exposures above the RfD as unacceptable (causing adverse effects) cannot be made (USEPA, 1991a).

c. Cumulative Risks and Hazards

The excess cancer risks or noncancer hazards, resulting from each chemical of interest for each route of exposure, are first calculated separately as described above. Next,

- The separate cancer risks are then summed for all chemicals and for all pathways applicable to the same potentially exposed population. The resulting value is the total excess cancer risk for that population.

- In contrast, noncancer hazards are additive only for chemicals that produce the same type of adverse effect (such as liver damage). Noncancer hazards are separately summed for all chemicals, exposure routes, and pathways applicable to the same population to obtain what is referred to as "Hazard Indices" for that population.

d. Baseline Risks

When determining the risks associated with environmental chemical contamination, the existing or "baseline" condition is typically evaluated. Baseline risks are those risks that might exist if no remediation (cleanup) or institutional controls (*e.g.*, property deed restrictions) were applied at a site (USEPA, 1989b). Two major purposes are served through this determination:

- The first purpose is to help assess if a site poses a current or potential risk to human health in the absence of any remedial action (USEPA, 1991b). By evaluating the baseline risks, it may be discovered that the site may present an imminent and substantial endangerment, for regulatory purposes. The baseline assessment also could show that risks are below regulatory action levels, meaning remediation is not needed.

- The second major purpose for determining baseline risks is to help determine remediation goals and objectives.

B. ECOLOGICAL RISK ASSESSMENT

As indicated previously, risk assessment is a process of collecting, organizing, analyzing, and presenting scientific data to inform decision-making processes. In ecological risk assessment, scientific methods are used to evaluate the likelihood of adverse ecological effects occurring because of stressors, such as hazardous substances or hazardous wastes. Although this assessment process is described in more detail in Appendix A, it is summarized here as a three-phase process framework (composed of Problem Formulation, Analysis, and Characterization; see Figure 3.4) with five steps.

1. Problem Formulation

As mentioned in §II.A.2 above, Problem Formulation is used at the very start to establish the purpose, objectives, and goals of the assessment. As the old saying goes, "begin with the end in mind" (Covey, 1989). There are several specific issues germane to formulating an ecological risk assessment.

1. There is the need to establish the ecological management goals of the process:
    o for specific things (*e.g.*, an animal, plant, or the water quality of a stream),
    o with definable identity attributes, and
    o have achievable states (whether they be in terms of locational occurrence, population size, physical/chemical state, or whatever other qualities are appropriate that reflect the societal values and needs of those involved).

2. Assessment Endpoints are "…explicit expressions of the actual environmental value that is to be protected" (USEPA, 1992a). They link the risk assessment to specific management concerns/goals. These endpoints identify:
    o a valued ecological entity or entities,
    o an attribute of that entity (or entities) which is important to protect and potentially at risk (*e.g.*, nesting and feeding success of an endangered bird species or the areal extent and patch size of an important/threatened plant), and
    o the spatial and temporal extent of interest.
    To be scientifically valid, assessment endpoints must be relevant to the ecosystem of interest and able to evaluate the stressors (contaminants) of concern.

3. Identify preliminary management alternatives for the entity and identify whether the risk assessment can actually support the achievement of these management alternatives. If it cannot help in this regard, then perhaps a risk assessment is not needed or different management goals and/or assessment endpoints need to be developed).

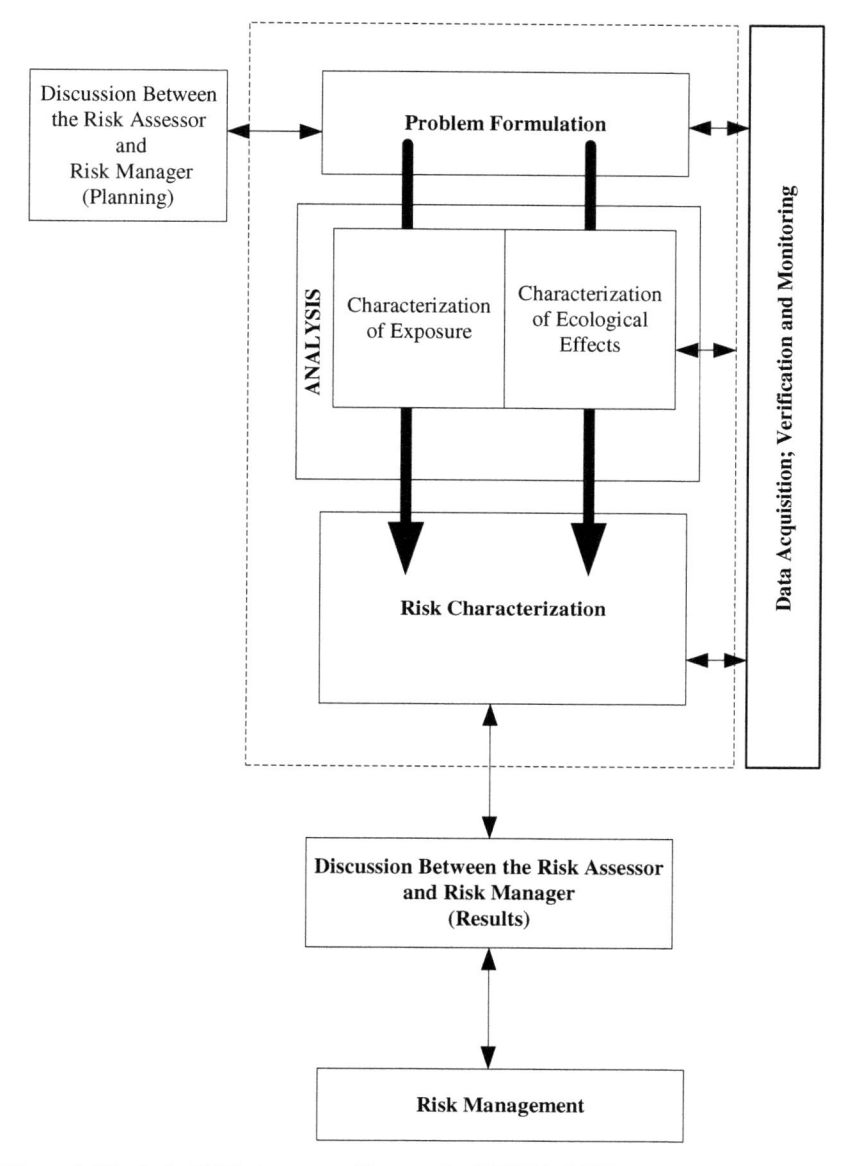

Figure 3.4 Ecological Risk Assessment Framework of USEPA (1992).

2. Hazard Identification

Hazard identification uses information concerning, for example, chemicals in the environment gathered during a site assessment to identify those things that may pose a risk to ecological resources, if exposure occurred. The qualitative exercise outlined below produces a list of potential hazards, *e.g.*, chemicals.

1. Site chemical concentrations are compared to available information concerning background levels, such as:
   - naturally occurring concentrations—ambient concentrations of chemicals present in the environment that have not been influenced by human activity; and
   - "anthropogenic" levels—concentrations of chemicals that are present in the environment originating from man-made, non-site sources (*e.g.*, industry, automobiles).

   The scientific literature is a useful source for background concentrations, but site-specific background data—that is, data collected adjacent to or off-site that have been unimpacted by site activity (*e.g.*, soil data collected "up-wind" of a site)—are preferable.

2. Site-wide frequency of detection—Chemicals with a frequency of detection of <5% in a large data set (n>19), or chemicals not detected at least once above the limit of detection, are generally excluded.

3. Site chemical concentrations may be compared to "screening" criteria or standards—As such, a chemical concentration greater than these criteria or standards simply indicates a need for more detailed evaluation.

3. Exposure Assessment

For ecological receptors, the total exposure is the sum of exposures from various components of the diet and from incidental direct contact with contaminated media (*e.g.*, soil, water, etc.). The cumulative dietary exposure is calculated by multiplying the tissue concentration in each prey item by the proportion of that prey item in the diet and adding these values. The total is then multiplied by the receptor's site use factor (SUF), exposure duration (ED), and ingestion rate (IR) and divided by the receptor's body weight (BW). As an equation it looks like this:

$$EE_{diet} = \sum \frac{(P_1 \, x \, T_1) + (P_2 \, x \, T_2) + ...(P_n \, x \, T_n) \, x \, SUF \, x \, ED \, x \, IR}{BW}$$

Where:

EE_{diet}	=	Estimated exposure from diet (mg/kg/day)
P_n	=	Percentage of diet represented by prey item ingested
T_n	=	Tissue concentration in prey item n (mg/kg dry weight)
SUF	=	Site use factor (unitless)
ED	=	Exposure duration (unitless), equal to the fraction of the year spent in the region
IR	=	Ingestion rate of receptor (kg/day in dry weight)
BW	=	Body weight of receptor (kg in fresh weight)

The SUF indicates the portion of an animal's home range that would be represented by the site. If the home range is larger than the site, the SUF is calculated by dividing the site area by the home range area. If the site area is greater than or equal to the home range, the SUF is set at one (1.0). If home ranges are not available for a particular species, the SUF defaults to one (1.0). The ED is the percentage of the year spent in the site area by the receptor species. For example, birds may be considered either year-round residents or migratory, with an ED set at 1.0 or 0.5, respectively.

The exposure of wildlife to chemicals through soil or water is estimated similarly to dietary exposure. The soil or water concentration is multiplied by the SUF, ED, and IR, and then divided by BW. The total exposure is the sum of exposure from diet, sediment ingestion, and surface water ingestion:

$$EE_{total} = EE_{diet} + EE_{soil} + EE_{water}$$

4. Toxicity Assessment

The toxicity assessment for ecological receptors describes the toxicological characteristics of agents of concern and establishes Toxicity Reference Values (TRVs) for each endpoint species identified at the site. These values represent NOAELs or LOAELs for each chemical and each endpoint species, and come from published toxicity (laboratory) studies. The species and conditions in a laboratory study often differ from those found in the field; therefore, some uncertainty is involved in extrapolating from the laboratory toxicity data to the TRVs. Just like human toxicity factors, because of this uncertainty, conservative factors are incorporated to assure protection of ecological resources.

5. Risk Characterization

In the risk characterization step, the toxicity and exposure assessments are integrated into quantitative expressions of risk. The risk from exposure to each individual chemical (or stress-causing agent) by each route of exposure, exposure pathway, category of receptor (*i.e.,* adult or child), and exposure case (RME or average) is first calculated. These separate, individual risks are then added across all chemicals and all pathways applicable to the same population to produce the total risk for that population. Potential risks are evaluated by calculating a Hazard Quotient (HQ) for each chemical and for each endpoint species. The HQ_{total} for all pathways was determined by dividing the total exposure via all pathways (EE_{total}) by the appropriate TRV for the endpoint species and contaminant:

$$HQ_{total} = \frac{EE_{total}}{TRV}$$

If the resulting HQ_{total} is greater than one (1.0), a potential risk for adverse effects from exposure to AOPCs may exist. The magnitude of the HQs is generally accepted as indicating a relative risk to the endpoint species under evaluation. By referring to the percentages of exposure resulting from different pathways (*e.g.,* food ingestion, soil ingestion, water ingestion), the relative contribution to total potential risk for each exposure pathway can be identified.

Besides providing an estimate of risk to the biological/ecological entities required for assessment, the assessor also must provide a description of the risk formulated along certain "lines-of-evidence." Agreement among the evidence about risk increases confidence in the conclusions of the assessment. These "lines-of-evidence" can include:

- demonstrating relevance of the measurement of risk to the assessment endpoints,

- demonstrating the relevance of the assessment to the conceptual risk system model,

- describing the sufficiency and quality of the data used in the assessment (for example, the strength of cause-effect relationship between the entity and chemical stressor and the magnitude and direction of uncertainties in the assessment), and

- determining the occurrence of ecological adversity (which is defined by the nature and intensity of those effects, the spatial and temporal scales of the effects, and the potential for recovery of the entity).

C. ACKNOWLEDGING UNCERTAINTY: THE RISE OF PROBABILISTIC RISK ASSESSMENT

Perhaps some wish risk assessment would be a strictly objective process, with decisions on risk based wholly on science. A considerable amount of uncertainty, however, is inherent in both human health and ecological protocols because of natural variability in sensitivity and behavior, uncertainty about exposure, and a lack of complete information about the toxicity of chemicals. Combining variability and uncertainty, with a limited database, means that determining the "true" risk associated with chemical exposure is unattainable. However, the pursuit of a "better" risk estimate is a worthwhile exercise prompting scientists, engineers, and policy analysts alike to explore the use of various quantitative uncertainty methods in risk assessment.

There is a growing push to include in every risk assessment a better explanation of the uncertainty and variability involved in the analysis:

- Provide quantitative estimates not as point values but as ranges, and verbally describe what they mean.

- Acknowledge the assessment's assumptions.

- If it serves the overall risk management strategy and will facilitate decision-making, then use probabilistic tools to quantitatively define the uncertainty and variability associated with the variables in the risk assessment.

Although Monte Carlo analysis (MCA), a probabilistic process, has seen some acceptance in environmental risk assessment, it has been used for years in finance and insurance (Rugen and Callahan, 1996). Probabilistic risk assessment (PRA) is a general term for risk assessments that use probability models to represent the likelihood of different risk levels in a population (*i.e.*, variability) or to characterize uncertainty in risk estimates.

- In human health risk assessments, probability distributions for risk reflect variability or uncertainty in exposure.

- In ecological risk assessment, risk distributions may reflect variability or uncertainty in exposure or toxicity.

An assessment that includes an analysis of variability can be used to address the following:

- "What is the likelihood (*i.e.*, probability) that risks to an exposed individual will exceed a regulatory level of concern?"
  - ○ For example, based on the best available information regarding exposure and toxicity, a risk assessor might conclude, "It is estimated that there is a 10% probability that an individual exposed under these circumstances has a risk exceeding 1×10^{-6}."
- If a probabilistic approach also quantifies uncertainty, the output from a PRA can provide a quantitative measure of the confidence in the risk estimate.
  - ○ For example, a risk assessor might conclude, "While the best estimate is that there is a 10% chance that risk exceeds 1×10^{-6}, I am reasonably certain (95% sure) that the chance is no greater than 20%."

USEPA has published risk assessment guidance setting forth the Agency's policy about the use of MCA, the conditions for acceptance of risk assessments using these techniques, and implementation (see discussion in §X of Appendix A and Figure 3.5). The principles include procedures for deciding the value of conducting such analyses, selecting input data and distributions, evaluating variability and uncertainty, running the calculations, the use of sensitivity analyses, and result presentation.

It is interesting to consider the use of these techniques for the informational purposes of risk characterization, that is, how do environmental managers, let alone risk assessors, inform the public with clarity about the meaning and use of risk distributions, as compared to point estimates? Anyone who has ever stood before the public knows that the only answer people are likely to accept to questions such as, *"Are my children and I safe?"* or *"I have cancer, did this site cause it?"* is *"yes"* or *"no."* As Alan Stern has observed,

> *"[PCCRARM] emphasized the importance of stakeholder involvement ... agencies are responding. [Simultaneously risk assessors are] ... increasingly convinced of the appropriateness of ... [MCA]. [As some in] the public are ... beginning to grasp the basics of risk assessment, we are preparing to put it beyond most people's reach."* (Stern, 1997)

The point is not to argue for or against the use of probabilistic risk assessment and MCA, but to emphasize that the use of these techniques must be considered:

- in the larger context of the goals and objectives of the environmental risk management effort,

- the value of the information content contained in the data provided, and

- the need to describe the risk system requiring control in the best manner possible.

D. INFORMATION DEVELOPMENT, NOT PROBLEM SOLVING

As defined in Chapter 1, Risk Analysis is a two-component process:

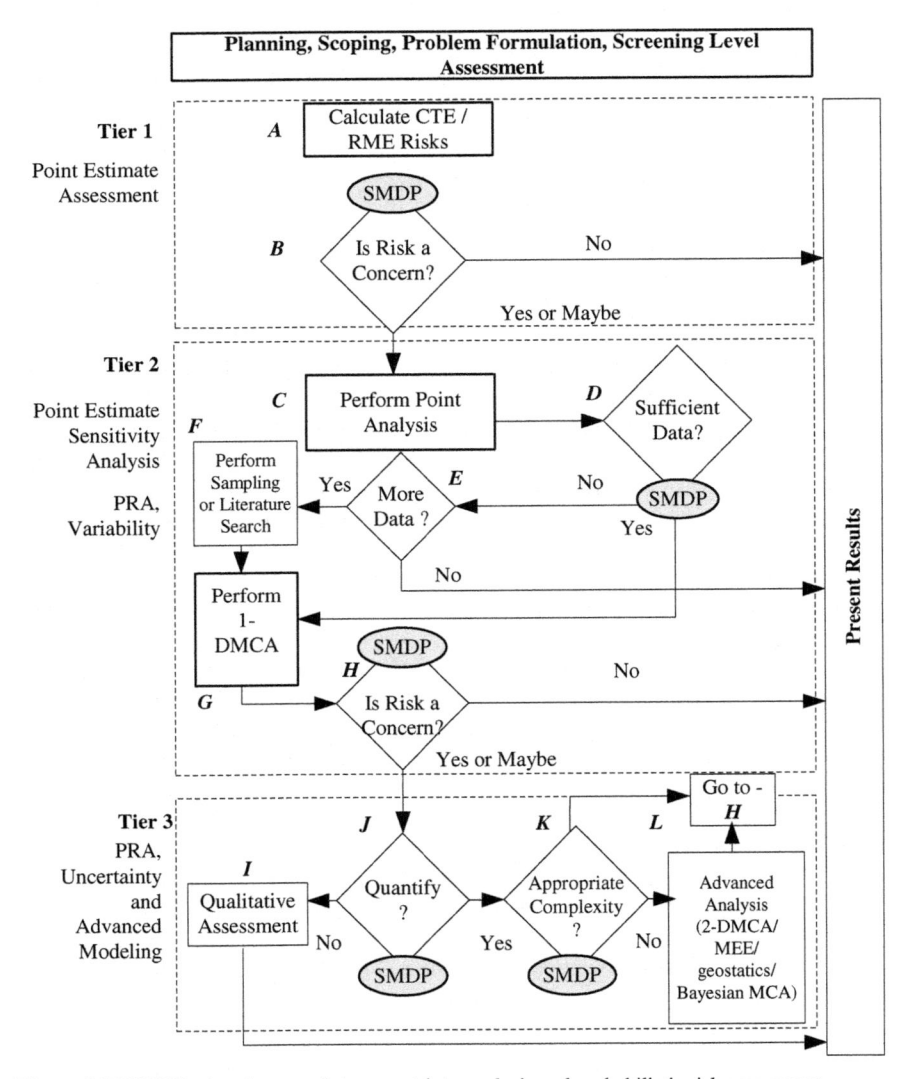

Figure 3.5 USEPA's tiered approach to uncertainty analysis and probabilistic risk assessment (after USEPA, 1999b). SMDP - Scientific / Management Decision Point, CTE - Central Tendency Exposure, MEE - Micro-Exposure Event Analysis, RME - Reasonable Maximum Exposure, 2-D MCA - 2-Dimensional Monte Carlo Analysis.

1. Evaluating (qualifying and quantifying) risk(s), generally considered being risk assessment.

2. Making (policy or reuse) decisions based on the foregoing evaluation together with other input (*i.e.*, the "whole picture"), generally considered risk management and risk communication.

Over the last decade or more, people have come to understand that there needs to be a closer, more constructive integration of these components. Risk assessment as generally practiced is a useful tool for information development, as we have seen, but it has no problem-solving power by itself. The discussion that follows, together with Figure 3.6, outlines what risk assessment provides to environmental risk management decision-making.

1. Use in Developing Environmental Site Assessment Plans

Early identification of exposure media, routes, and points are beneficial when designing site investigation and sampling plans to ensure that the appropriate number, type, and location of samples needed to assess exposure are taken. If some results are available, the use of risk-based screening techniques (*e.g.*, USEPA's soil screening guidance, ASTM RBCA Tier 1 screening levels) can help focus subsequent investigations and help the manager begin to appreciate potential issues.

2. What Risk Assessment Does and Does Not Tell You

Risk assessment provides quantitative estimates of current or potential future risks associated with chemical and other contaminants in the environment. This operational definition represents a fair statement about what it does. However, there are limitations and a number of things that it does not or cannot tell us.

- **Estimates**—Any assertion of preciseness of a risk value, or that calculated risk values are measures of true or actual risk, is false. This fact is hard for many to understand let alone accept. The whole protocol dictates estimates only, and overestimates at that, *"...USEPA is reasonably confident that the 'true risk' will not exceed the risk estimate derived through the use of this model and is likely to be less than predicted"* (USEPA, 1989a).

- **Population Risks**—Assessments tell you nothing about *individual* risks.
  - They provide probabilities that a certain number of persons within a given population will contract cancer over the course of their lifetimes. For example, a cancer risk of 1×10^{-6} represents one cancer in excess of background in a population of one million persons. A cancer risk of 1×10^{-6} does not mean that any individual person has a one-in-one million chance of contracting cancer. There are means to calculate individual risks, but they are not typically employed in most risk assessments.
  - Critics claim that risk assessment works best to describe small risks to large populations, but that it does a poor job of telling us about large risks to smaller groups, such as workers, the economically disadvantaged, or ethnic minorities. Opponents also argue that the science used in risk assessment is immature and suitable only for evaluating immediate threats or the risk of developing cancer, or to focus on problems or aspects of problems that already are well

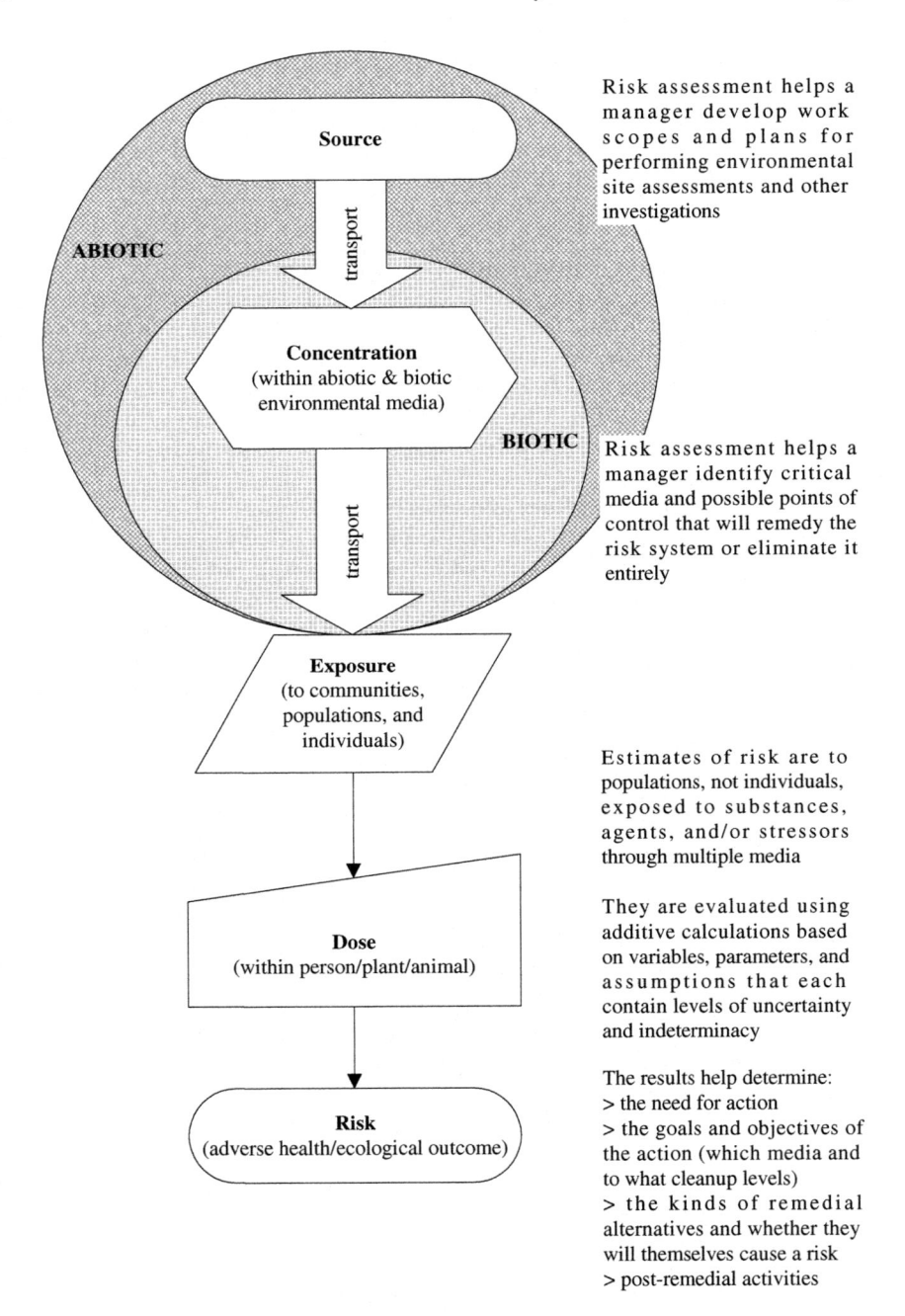

Figure 3.6 What risk assessment tells you about the risk system and its control.

understood. While it is not within the scope of this chapter to address these criticisms in detail, it is sufficient to say here that assessments can be constructed to successfully address these issues.

- **Mixtures**—Risk assessments for most sites rarely account for potential chemical interactions like synergism and antagonism, even though most people generally are not exposed to a single chemical but mixtures. To account for this fact, the protocols assume additivity approach; that is, exposures and risks are summed (added) across all chemicals detected at a site. This approach, however, fails to account for the possibility that exposure to two or more chemicals may produce effects that are greater than additive (*i.e.,* synergistic), or that the effects may be less (*i.e.,* antagonistic). Nevertheless, contrary to many people's fears, synergistic or antagonistic interactions rarely occur in exposures to low levels of chemicals in the environment. This is an important justification for the general practice of assuming additivity.

- **Uncertainty**—Despite appeals on many fronts to include at least qualitative descriptions of uncertainty in risk estimates, most assessments do a poor job. The advent of probabilistic techniques (see above) and USEPA's publication of draft PRA guidance (USEPA, 1999b) should address this shortcoming and provide valuable information and better understanding, if assessors do it and managers ask for it.

Clearly, characterization of risk should help regulators, stakeholders, and the public grasp the context of the situation, in that we are dealing with estimates of what is possible. The risk assessment is not factual evidence of exposure (or even risk), but only *"grist for the decision making mill."* This decision-making involves public health and not individual, high-end risks. Thus, the risk characterization should carefully articulate that the risk estimates presented in the report represent finite, incremental increases to existing background exposures.

- Instead of only providing the following aspects:
  - define the potential maximum "individual receptor" incremental exposures,
  - characterize the health risks associated with the "individual receptor" incremental exposure, and
  - base risk management decisions on the acceptability of the potential incremental risks. If risks are "unacceptable," clean the site to the background environmental level, or to an acceptable level, whichever is highest.
- The risk assessment should:
  - define average background exposures to substances of concern at the local, regional, and/or national level,
  - define additional "individual receptor" exposures due to the site,
  - characterize the health risks associated with these background exposures,
  - establish priority classes using the risk estimates for background exposures, and

○   base risk management decisions on the priority classes of the
substances of concern and the potential added exposure from the site.

3. Use in Determining the Need for Remediation or Corrective Action

As discussed earlier, baseline risk assessments evaluate the current and expected
future risks posed by a site in the absence of cleanup, taking into account expected
land use and employing assumptions about exposure that are conservative. These
results help define whether either human health or ecological risks are significant
and justify the need for cleanup to reduce those risks at the site. Risk assessment
helps managers identify principal threats. USEPA defines these threats in part as
materials

> *"...with toxicity and mobility characteristics that combine to pose a*
> *potential risk several orders of magnitude greater than the risk that*
> *is acceptable for the current or reasonably expected future land use,*
> *given realistic exposure scenarios."* (USEPA, 1997b)

4. Use in Developing Cleanup Goals

If cleanup is deemed necessary at a site, cleanup criteria are identified that
represent an "acceptable" level of residual concentrations in the affected
environmental media. Cleanup levels can be established by one (or a combination)
of the following procedures (Washburn and Edelmann, 1999):

- relying on environmental quality standards from applicable or relevant
  statutes (*e.g.,* federal or state drinking-water standards);

- using local background concentrations or analytical detection limits; and

- using risk assessment methods to determine concentrations that are
  protective of human health and the environment, given reasonably expected
  land use or institutional controls that might be considered.

Risk-based cleanup levels are developed using the basic methodologies and
assumptions applied in the baseline risk assessment. Cleanup levels are "back-
calculated" to correspond to an overall risk goal for the site. This goal is
accomplished by using a pre-determined level of risk and solving the exposure and
risk equations from the baseline risk assessment for the concentration term of the
equation.

A recent paper by Schulz and Griffin (2001) argues that preliminary remediation
goals (or PRGs) developed by USEPA and many others are used inconsistently at
Superfund and other sites. Often they are treated in remediation decisions as "not-
to-exceed" concentrations (or NTECs). When used in this manner, overly
conservative and unnecessarily expensive cleanup actions result. When applied
correctly and according to their original intent, PRGs and other typical cleanup goals
are merely a starting point around which statistical methods or a simple iterative hot
spot removal approach is used to develop appropriate NTECs. These NTECs ensure
that average post-remediation concentrations for a particular contaminant are at or
below the PRG. Although the NTEC is a concentration set higher than the PRG, it is
nevertheless protective of human health and the environment when applied to the
relevant area of the site under consideration. Therefore, because NTECs result in a

higher concentration, the related cleanup costs can be substantially less than those based on concentrations set equal to PRGs.

Risk-based methods provide a scientifically defensible and standardized way for determining cleanup goals for a site. However, as discussed in the last section (§V.C) of Chapter 2, the need to cleanup and the establishment of cleanup goals is not a simple exercise of "just give me the numbers!" Environmental risk management should not be driven by concentrations in the environment but by the potential risk. We will get back to this point later.

5. Use in Selecting Potential Remedial Alternatives

If remedial action is necessary, remedial alternatives must be evaluated and an appropriate solution chosen, designed, and implemented. First, the chosen alternative must meet two threshold criteria:

- regulatory compliance with legally applicable or relevant and appropriate federal and state requirements, standards, criteria, or limitations (that is, ARARs), and

- the solution also must protect human health and the environment.

While the focus here is not on ARARs, it is important to remember that there are three types of ARARs:

- Chemical-specific (which limit specific hazardous substances in the environment)

- Location-specific (which restrict activities based on site characteristics and conditions)

- Action-specific (which are technology-based restrictions caused by certain remedial actions)

If there are no chemical-specific ARARs or if they do not define sufficiently protective levels, then risk assessment is the tool by which one develops cleanup levels that will protect human health and the environment, as discussed earlier.

A remedial procedure is selected to attain the cleanup goals. The selection of the best three or four procedures involves balancing five criteria:

- long-term effectiveness and permanence,

- reduction in the toxicity, mobility, and/or volume of the waste or substance(s),

- short-term effects,

- implementability, and

- cost-effectiveness.

The surviving alternatives are judged by two more criteria:

- state's acceptance, and

- community acceptance.

Risk assessment methods can shed light on all of these considerations. In particular, long-term effectiveness is most frequently based on an evaluation of the residual risk posed by a site after remediation is complete. Evaluation of short-term effectiveness often includes an assessment of risks of implementing a remedy, such as acute risks associated with the excavation of waste materials.

6.   Use in Evaluating the Risk of Remedy

As discussed, risks associated with the implementation of remedial alternatives at a site are considered as part of the evaluation of short-term effectiveness. Various statutes require consideration, during remedy selection, of the risks of implementation at hazardous waste sites. USEPA (1998b) guidance states, "*...alternatives should be evaluated with respect to their effects on human health and the environment during implementation of remedial action.*" The evaluation should include, "*...any risk that results from implementation of the proposed remedial action...that may affect human health...and threats that may be posed to workers.*" Such risks include: exposures to chemicals, accidents associated with the use of heavy equipment, heat stress caused by impermeable protective clothing and use of respirators, and accidents and spills during transportation of hazardous materials (Washburn and Edelmann, 1999). Populations potentially at risk during implementation include on-site workers during investigations and cleanup, off-site residents, workers in nearby areas, and crops, livestock, and wildlife near the site.

7.   Use in Planning of Post-Remediation Activities

The nature of risks posed by a site should be considered during development of post-remediation monitoring plans (Washburn and Edelmann, 1999). For example, a groundwater monitoring plan developed in response to short-term risks from concentrations that may fluctuate above a certain threshold will likely require more frequent sampling events than one driven by cumulative risk over many years of exposure. Additional risk-related factors to consider when identifying the sampling frequency requirements in a groundwater-monitoring program include proximity to down-gradient receptors, the groundwater flow rate, and whether or not seasonal changes occur in the groundwater aquifer. Risk assessment can also be used to narrow the focus of each groundwater-sampling event so that analyses focus only on the contaminants that contribute most to overall risk.

Risk assessment can also assist in evaluating the need for institutional controls as well as the controls that may be appropriate for a particular site. With contaminated sites such as those addressed by USEPA in its Superfund program, the core idea of Institutional Controls is to prevent risk to human health by preventing *exposure* to contaminants (English *et al.*, 1997). Risk assessment can identify site uses that are incompatible with current conditions but, with the use of certain institutional controls, may be appropriate for other (less restrictive) uses. For example, controls such as deed restrictions and zoning changes may permit use of a site for commercial purposes after the risk assessment indicates that residential use would result in the existence of a potential health threat. A site for which long-term groundwater consumption poses a health risk could remain viable if institutional controls such as restrictions on well-digging, well water testing, or alternative water supplies are considered.

III. RISK MANAGEMENT: THE JOB EVERYONE WANTS

Because a number of stakeholders may have opinions about how best to manage risks at an environmentally impaired site, sorting out who "the" risk manager is, can sometimes be a tricky endeavor. Regulatory requirements generally dictate that the agency ultimately makes the final call concerning the management of risk. Despite a seemingly concrete approach with the regulator serving as decision-maker, the corporate manager has considerable power. Other stakeholders likewise influence the process and often are not content to let the agency dictate the process.

To complicate matters, as we have said before, most risk assessors, engineers, and other environmental scientists tend to think only about the object to be produced – *i.e.*, "their" reports, a risk assessment for example. In contrast, the users (the risk managers and stakeholders) think about what the "risk assessment" will do for or to them. Such cognitive insularity results in tension and stagnation, and this insularity cuts the risk assessor off from having something in common with other stakeholders (*i.e.*, thought shapes reality).

A. THE REGULATOR'S JOB

As already discussed, regulatory agencies such as the USEPA typically dictate what levels of risk are high enough to warrant remedial action at a site. In the early years of the Superfund program, for example, a cancer risk of 1×10^{-6} (a one in one million cancer risk) was often informally used as a "point of departure" with respect to remedial decisions—that is, risks lower were considered negligible, but risks higher were presumed eligible for remediation (Walker *et al.*, 1994). Specifically, USEPA's policy (that most State agencies follow) for making decisions concerning exposure to known or suspected carcinogens:

> "*For known or suspected carcinogens, acceptable exposure levels are generally concentration levels that represent an excess upper bound life-time cancer risk to an individual of between 10^{-4} and 10^{-6} using information on the relationship between dose and response. The 10^{-6} risk level shall be used as the point of departure for determining remediation goals for alternatives when ARARs are not available or are not sufficiently protective because of the presence of multiple contaminants at a site or multiple pathways of exposure....*"
> (40 CFR 300.430[E][2][i][A] [2], USEPA 2001a)

The intent was that decisions concerning remediation would be flexible when risks fell within this range. The results of a baseline risk assessment provide important information about what is or is not a significant exposure and/or risk and, therefore, what is or is not of concern, based on existing regulatory requirements. To better explain this issue, consider the discussion in Table 3.1.

The Harvard Center for Risk Analysis notes that numerical risk management criteria such as these, although widely used in federal and state environmental programs, are rarely mandated by statute, often change over time, and have rarely been set based on careful policy analysis (Walker *et al.*, 1994). Furthermore, in contrast to the general belief that the National Contingency Plan and USEPA have a clearly stated preference for managing risks at the more protective end of the risk

Table 3.1
Carcinogenic Risks

Let us say that a population group living in the U.S. has a chance of contracting cancer (from all sources except the contaminated site) that is 1 to 1. If that group is exposed to a site as described in this risk assessment, the decision making about potential cancer risks can be divided into three parts as shown below.

If the site increases the exposed population's chance of contracting cancer to	1.0001 to 1 or more	*Then this exposure requires reduction.*
If the site increases the exposed population's chance of contracting cancer to between	1.0001 to 1 and 1.000001 to 1	*Then this is in a range at which reduction in exposure may be required.*
If the site increases the exposed population's chance of contracting cancer to	1.000001 to 1 or less	*Then this does not require reduction in exposure.*

range (*i.e.*, 1×10^{-6}), it is important to note a paper by Doty and Travis (1989). These authors reviewed fifty Superfund Records of Decision and concluded that risk management decisions were predominantly formulated on the basis of the *less* protective end of the risk range (*i.e.*, 1×10^{-4}). Of the twenty-three sites (out of 50) for which remediation was based on current risk, overall site risks were at 1×10^{-4} or greater for thirteen of them (or 56%). Only four of the sites with risks of 1×10^{-6} or less were slated for remediation (or 17%). Likewise, of the twenty-five sites for which remediation was based on future risk, site risks were 1×10^{-4} or greater for all but two (or 92%).

USEPA's policy (that most state agencies follow) for making decisions concerning exposure to noncarcinogenic chemicals (or systemic toxicants) can be stated as:

> *"For systemic toxicants, acceptable exposure levels shall represent concentration levels to which the human population, including sensitive subgroups, may be exposed without adverse effect during a lifetime or part of a lifetime, incorporating an adequate margin of safety...."* (40CFR 300.430[E][2][i][A][1], USEPA, 2001a)

Exposures that are less than the RfD—*i.e.*, exposures with a noncancer hazard value less than one—will not be associated with health risks. Exposures exceeding the RfD—*i.e.*, exposures with a noncancer hazard value greater than one—may be associated with adverse health effects in a population. Nonetheless, a clear distinction that would categorize all exposures below the RfD as acceptable (*i.e.*, risk-free) and all exposures above the RfD as unacceptable (causing adverse effects) cannot be made (USEPA, 1991c). With regard to what is or is not a significant exposure and/or risk, consider the discussion in Table 3.2.

Table 3.2
Noncarcinogenic Risks

Acceptable exposure levels are those to which people, including those who may be more sensitive (children, for example), may be exposed without adverse effects during their lifetime or part of their lifetime. This acceptable level also includes an additional margin to ensure safety. This level is called a reference dose and is, in essence, a safe dose.

If the exposure level is *below* *this* acceptable level	*Then it does not require reduction.*
If the exposure level is *above* this acceptable level	*Then it is in a range at which reduction in exposure may be required.*

B. THE VIEW OF OTHERS

The business or corporate perspective can be characterized as one of a desire to understand what is required, what the process is, where the latitude is, and then to work through that process at a pace and expense dictated by their business calculus.

From the stakeholder perspective, they seek validation of their concerns and assurance that their health, the public's health, and the environment will be protected, among other things (refer back to §V.B.2 of Chapter 2). Additionally, as discussed in Appendix C, a few stakeholders will agree with whatever the plan is, and a few others will disagree and hate whatever decision is made. Because the benefits and costs being weighed are social by their nature, people's values and preferences are involved making the decisions at least apparently political (Toll, 1999 and Vincent, 1999).

C. THE RISK MANAGER IS A PROCESS, NOT A PERSON

As mentioned in Chapter 2, Portney (1991) identifies two principal styles of public environmental decision-making: positivism and public policy. Yes, there are various regulatory guidelines, standard protocols, and the like directing that regulators must discharge their public duty accordingly. Yes, they will from time to time fall into the trap of the status quo seeking to avoid decisions that are innovative or set precedent. Nevertheless, because the problem is one of resolving conflicting worldviews, ambiguity, uncertainty, and objectives, the behavior and decisions of the regulatory organizations charged with the decision on behalf of their constituency cannot be solely positivist (Roome, 2001). Some see this as a bad thing, as suggested by Charnley's (2000) findings. We think it should be seen in a positive light.

The upshot is that no one organization or person is the risk manager. Rather, the risk manager is a network of multiple decision-influencing and decision-making risk managers, including site neighbors, community leaders, other stakeholders, the regulated entity, and the regulators who work in this process. This network means there are multiple decision objectives, multiple frames, multiple bases of knowledge, and many opinions. Understanding the science, the legal and regulatory requirements, and the politics, and being armed with and executing a coherent management strategy and communications plan (that has a strong educational component, Appendix C) means that the manager has a good chance of achieving a favorable risk management decision.

IV.   RISK COMMUNICATION: THE JOB NOBODY WANTS

A.   NECESSARY AND DIFFICULT

Risk communications is the exchange of information about health and environmental risks among risk assessors, risk managers, the public, media, interested groups, and others. This definition suggests that this communication is not one-way but rather an interactive process involving the transfer of risk messages, the expression of reactions, concerns, and opinions, and listening. It is not necessarily an easy or smooth process, and there will be conflict (Gorczynski, 1992). This situation is demonstrated if one considers four common, yet erroneous, assumptions about this type of communication (Wolfe, 1993):

- the best and only proper approach to the evaluation of risk is through quantitative, probabilistic means;

- high-quality, tailored presentations of risk messages will convince the public;

- agreement with expert risk messages and a quantitative evaluation will yield similar agreement to follow-on risk management plans; and

- risk communication is a discrete, formal process.

Wolfe critically demonstrates that each of these assumptions is flawed, and goes on to show:

- the importance of appreciating the social context of the situation (remember the discussion of risk importance, risk significance, risk as harm, and the matter of uncertainty at the end of Chapter 2);

- that the risk communication is interactive; and

- that risk communication is an ongoing, continual process.

Although we might like to make all of the decisions privately and then proceed, the Decide-Announce-Defend (or DAD) approach will likely cause more trouble than it is worth. The DAD approach is often favored because it provides the feeling of control and the sense that the decision-making process is inexpensive (Robinson, 2000). However, there is another way that incorporates crucial elements to meeting the challenge incumbent in the risk communication process, coping with the feelings of loss of control, and coping with the financial uncertainties of the process.

B.   MEETING THE CHALLENGE

Much has and can be said about this topic; to this end, Appendix C describes the associated "Rules" and Priorities of Public Communication, Decision Priorities, and the development of a communications strategy and plan when involved with managing impaired properties and their remediation.

Recently, using Fischhoff's (1995) speculative, evolutionary account, Chess (2001) describes a series of communication strategies highlighting the development of risk communication, and then proceeds, using the tragedy in Bhopal, to suggest that organizational theory adds a deeper understanding of this development. The paper makes a case for seeing risk communication as arising out of the interaction of corporations within their organizational environments, struggling to achieve or maintain legitimacy as required by that environment (composed for example by stakeholders and other corporations demonstrating skillful risk communication approaches). She also makes a statement,

> *"[r]isk communication is not merely a response to the external environment; it also creates that environment,"*

which is similar to that of Tillich (1967) cited in §II.B of Chapter 1

> *"[r]eality precedes thought; it is equally true, however, that thought shapes reality."*

Fischhoff (1995) suggests that the current state of development of risk communication is exemplified by the following behaviors: corporations are getting the right numbers, explaining what they mean with historical perspective, showing why they are good, being nice to all, and making people partners with them. This is somewhat similar to the approach description of *Define-Agree-Implement*, or "DAI," that requires gathering identified key stakeholders to explore concerns in order to develop practical solution pathways to resolving an issue, as well as the system known as "fair process." The systemic study of "fair process" began in the mid-1970's when two social scientists, John W. Thibaut and Laurens Walker, combined their interest in the psychology of justice with the study of process (Chan Kim and Mauborgne, 1997). The research of these two scientists revealed that people are as interested about the fairness of the process through which an outcome is achieved as they are about the outcome itself. Subsequent study by Chan Kim and Mauborgne demonstrated that individuals are most likely to trust and cooperate freely with systems—whether they win or lose by those systems—when fair process is observed. A common belief is that people are concerned only with what is best for them; however, there is ample evidence to suggest that when a process is believed to be fair, most people will accept outcomes that are not totally in their favor.

"Fair process" is generally described as having three elements (Chan Kim and Mauborgne, 1997):

- Engagement—involves individuals in the decisions that affect them by asking for their input and allowing them to refute the merits of one another's assumptions. It communicates respect for individuals and their ideas. Encouraging refutation sharpens everyone's thinking and builds better collective wisdom.

- Explanation—means that everyone involved and affected should understand why final decisions are made the way they are. Explanation of the thinking underlying decisions aids people's confidence that their opinions were considered and that decisions have been made in the best interests of all involved. An explanation builds trust even if anyone's own ideas are rejected.

- Expectation Clarity—requires that once a decision is made, the new "rules-of-the-game" be clearly stated. To achieve "fair process," it matters less what the new rules are and more that they are clearly understood.

This process is not decision by consensus; it does not seek harmony or win people's support through compromises that accommodate everyone's opinions, needs, or interests. While it gives every idea a chance, the merit of those ideas rather than consensus drives decision-making. It pursues the best ideas whether articulated by the few or the many.

The "fair process" of Chan Kim and Mauborgne is similar to the Analytic-Deliberative process of Renn (1999). Renn's "cooperative discourse" model has stakeholders, experts, and randomly selected citizens work through three respective steps: value and criteria elicitation, development of performance profiles of each policy option, and evaluation and design of the policies. The sequential involvement of these groups, as shown in case studies, demonstrates that analytic thinking and deliberative argumentation must be integrated and not separated from decision-making to ensure fairness and efficiency.

Along these same lines, an effective form of interfacing, one that fosters problem solving, is known as the "mutual-gains" approach. Table 3.3 outlines the six simple guidelines that define this approach (Susskind and Field, 1996).

The corporate manager may be tempted to adopt only some of the mutual-gains approach, but each guideline is related to and informs the others. Following some but not all of the guidelines reduces their overall effectiveness. Ignoring one guideline or another may lead to actions that may be viewed as contradictory and thus may accentuate, rather than reduce, community concerns. However, the mutual-gains approach should not be applied in a blind fashion, without taking stock of each situation's unique aspects. The approach is a guide at best that informs experienced judgment; it is not a cookbook applicable in every instance.

Finally, as discussed in Appendix C, successful risk communications programs reflect several key elements:

- commitment of senior management;

- communications that are thoroughly planned and directly involve all elements of the community in the environmental remediation process;

- attention that is constantly focused on the fact that an impaired property, the public discussion of issues related to it, and any remediation are subject of an understandable management process; and

- a management process that is always focused on addressing the priorities the public has been trained to expect.

V. MANAGING THE INTERFACE

A. ENVIRONMENTAL NEGOTIATIONS

So far, we have quickly reviewed the paradigmatic troika of risk assessment—risk management—risk communication as separate aspects, but as often hinted in the discussion, they are not distinct but cohesive, overlapping aspects of a process. As such, it is crucial that these aspects effectively work together to ensure an efficient, legitimate process (Lopez Cerezo, 1999). This process can be thought of as a game, an "environmental negotiations" game, which was defined by Gorczynski (1992) as

Table 3.3
The Mutual Gains Approach

Acknowledge Concerns of Others	Essentially, "stand in another's shoes." It is in this way that one can begin to appreciate other's gestalts (*i.e.*, thought shapes reality). In so doing we can avoid the "zero sum" game of conflict perpetuation. If each side can explain the viewpoint of the other, it increases the likelihood of collaborative solution.
Encourage Joint Fact-Finding	The goal here is the generation of information believable by all. This may be difficult in the corporate setting, seemingly against the advice of legal counsel and potentially compromising proprietary information. The corporate manager wants to have the best possible information to be certain they are making the best decision, but the "best possible" information may not be the most convincing. Thus, the manager must decide what information others will find most useful. Information gathered behind closed doors may have little or no credibility when it sees the light of day, even if it is the "best information." A better approach is to open the doors wide and pursue joint fact-finding. In other words, gather and analyze data and draw conclusions together. This may be an unpalatable proposition for someone who wants to control the outcome, but with a skeptical public, joint fact-finding is far more likely to lead to believable outcomes.
Offer Contingent Commitments to Minimize Impacts if They Do Occur, Promise to Compensate Knowable but Unintended Impacts	Again, straightforward, but difficult because of what it means to management from a legal perspective. Nevertheless, if the company promises that something will not or cannot happen, it should stand behind that promise with an offer of contingent compensation.
Accept Responsibility Admit Mistakes Share Power	While not complicated, this may be difficult for a corporation. Mistakes and accidents happen, but the natural human reaction is to avoid blame in order to avoid liability. Yet, the liability exists regardless of disclosure, it's only a matter of time until disclosure occurs and the apparent and/or real costs to the corporation may ultimately prove insurmountable. Power sharing, allowing "non-experts" to assist, also can be distasteful. However, there is much to be gained in terms of goodwill relations by doing so.
Be Trustworthy, Always	This guideline is closely aligned with the previous one. The concept of trust is debatable, but methods of trust building are relatively concrete. To inspire trust, shape expectations and then follow through: "say what you mean and mean what you say." When it comes to trust, the old saying still applies, "use it or lose it."
Build Long-Term Relationships	This guideline may be most vexing for the corporate manager. Like all natural systems, relationships suffer from entropy, which means they tend to increase in disorder and chaos over time. The only way to avoid this trend is to expend energy to keep the disorder and chaos at an acceptable level. Many concerns vie for the attention of the corporate manager; thus building long-term relationships, especially with potentially adversarial community groups or other stakeholders, may not be on the top of the "to do" list. Nevertheless, if you as a manager are concerned about your company's image, its credibility, its future bottom line, you must focus on long-term relationships.

a serious process through which we decide the quality of the environment and life around us. This process has distinct periods that overlap and occasionally run concurrently, but generally flow in time sequence as follows:

- Discovery
- Initial assessment
- Decision for additional work
- Additional assessment phases, including *risk assessment*
- Development of options to control the risk
- *Risk management* decision-making concerning that risk
- *Risk communication*, public disclosure of decision and allowance for comment (as discussed above, communication occurs throughout the process)
- Formal engineering of the option
- Implementation of the risk management option
- Monitoring
- Closure

Gorczynski (1992) provides a typology of the players involved in the process:

- Engineers and other cool, dispassionate, scientific types
- Politicians: elected and otherwise
- Bureaucrats
- Industrialists and Developers
- Environmental Activists
- Interested People, the Public
- The Media
- Lawyers, Lobbyists, and Other Hired Guns
- Translators, Primary Leaders, and Bridgebuilders

Following Madsen and Ulhoi (2001), these players have a *stake* in this process game. Using the definition of Carroll (1993), a *stake* is: an interest, a legal or moral right, and/or ownership position.

Besides interacting with regulators, companies are learning that their environmental management systems must be capable of coping with dynamic networks of stakeholders, "...*driven by shifting meanings attributed to the environmental concerns of those stakeholders*" (Roome, 2001, Madsen and Ulhoi 2001). Along these same lines, Chess (2001) describes corporations as being part of a complex "organizational ecology web," which has as its home the urban-industrial ecosystem (mentioned in §I.A of Chapter 1 and presented in Figure 1.1), and which is where the impaired property of interest is located.

B. KNOWLEDGE-BASED NETWORKS

As suggested earlier, this network of players or stakeholders, including regulators and the company itself, is a type of knowledge-based network. The knowledge-based network can be defined as: a combination/linkage of loosely coupled and differentiated units—knowledge groups—that are interdependently related in the pursuit of an objective that no single unit can achieve by itself.

This definition is important because it recognizes different kinds of knowledge, objective and subjective, and it recognizes that the goal is mutual. As mentioned in Chapter 1, sustainable development can be defined as a coordinated action among business, government, communities, and individuals that leads to meeting present needs without compromising the ability of those in the future to meet their own needs (Robinson, 2000). The company's efforts are essential here because they have the financial and human resources, reach, and technology to redesign the environmental risk system (Figure 1.2) in a way that ameliorates environmental liability, elevates the market value of a property, and restores the property to improving the community's abiotic and biotic structure and cultural, socioeconomic function (Figure 1.1).

As we have seen, the corporate manager must effectively link and interact (network) with internal (Figure 2.8) and external knowledge groups (Figure 2.10), cope with the increased need for organizational flexibility and timely responsiveness, to deal with the risk that the regulators and stakeholders want managed. To this end, the corporate manager needs to execute a classic risk management process (recall Figure 2.6), which necessarily involves risk assessment and risk communications. However, as early as possible, before the process proceeds too far, the corporate manager needs to establish a strategy seeking to:

- manage the definition of the risk,

- engage and interface important and interested stakeholders as critical knowledge groups, so as to educate and learn,

- explain the risk system and its management and demonstrate an appreciation of differing worldviews,

- ensure clarity of expectations concerning the process and decision-making, and

- seek mutual gains for
  - the company, by reducing its financial and legal risks at the lowest possible investment of time and money;
  - the public, by protecting them and improving the community in which they live;
  - the regulators, by protecting human health and the environment; and
  - other stakeholders, by meeting mutually accepted goals.

C. THE VALUE OF RISK-BASED ANALYSIS

Environmental impairment leads to risk—environmentally and financially—and the real problem is how to cope in a (financially and reputationally) successful way with that risk, hopefully to reap some rewards or at least avoid significant losses. The best place to start the process is at the beginning, and that is where the Risk-Based Analysis comes into play.

As shown in Figure 2.10 the company's definition of risk and approach to managing it goes through not only regulatory scrutiny but also public-political scrutiny in order to achieve an external environmental risk and recovery management decision. Environmental risk management considers not just hard, objective, scientific knowledge (or brute facts, observation, and evidence as discussed in Chapters 3 and 5 of Conces, 1997) but thoughtful integration of subjective knowledge (or ideology-laden facts, observation, and evidence as discussed in Chapters 4 and 6 of Conces, 1997). As stated earlier, although environmental impairment may be on my company's property, the evaluation of "loss exposure" takes on an entirely different perspective when others (stakeholders) think about situations involving their exposure (to the property's "impairment") and their loss (of welfare, health, life, or just impairment of their "quality of life"). We believe that a management process using this currency of knowledge, that appreciates that "evidence" can and is ideologically laden can help depolarize the process, and then seeking mutual gains will be the most successful.

The role of Risk-Based Analysis is to place the impairment (that is, the contamination constraint) in perspective (*i.e.*, it helps the manager observe it through science and technology and see it through the eyes of other stakeholders). In so doing, it points the way to cost-effective management solutions by interacting with internal company planning and financial management embedded within an overarching legal framework (Figure 3.7). The objective is to manage the remedial process like a redevelopment process by strategically guiding environmental engineering to apply effective remedial tools in a coordinated fashion with planning, site/civil, transportation, and other activities to achieve exposure (and risk) mitigation within a context where *redevelopment is remediation.* **Thus, the application of Risk-Based Analysis is not to justify less cleanup but build a financially sound approach that people (stakeholders) will agree as being safe**. The purpose is to seek mutual gains (Susskind and Field, 1996).

Risk-Based Analysis will be useful to the corporate manager in the initial development of a management and communications strategy for an environmentally impaired property and its remediation (Figure 3.7). It is a tool for information development and the gleaning of insight into the objective and subjective aspects of an environmental impairment. It helps the manager develop a clear and understandable definition of the risk system that appreciates other's goals and values, through the characterization of not only the environment, but also the socio-political, socioeconomic, legal/regulatory, and financial aspects as well. Risk-Based Analysis is not "risk assessment," although risk assessment is a part. Rather, it is a process to help the corporate environmental manager better implement, direct, and use risk assessment, risk communication, and influence the multi-component decision process of risk management. Chapter 4 describes Risk-Based Analysis and shows how it functions to inform the environmental risk management process throughout its operation.

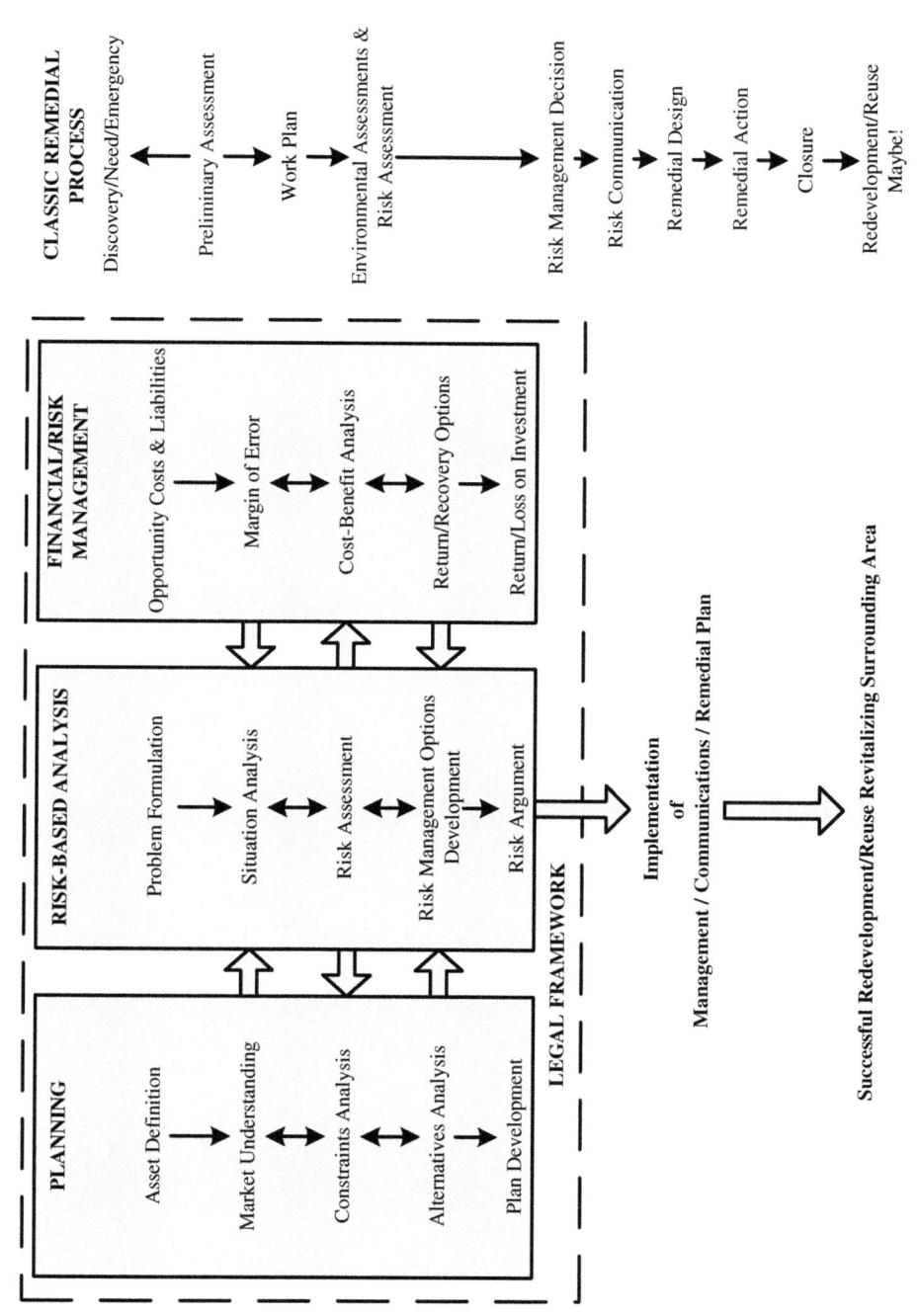

Figure 3.7 A holistic approach to interfacing the assessment, management, and communication of risk.

Chapter 4

THE PRACTICE OF RISK-BASED ANALYSIS

Kurt A. Frantzen, Cris Williams, Judy Vangalio, and Jerry Ackerman

I. INTRODUCTION

Harte (1988) suggests a three-step approach to environmental problem-solving:

- Develop a broad overview of the problem to obtain "the big picture" while applying appropriate reality checks.

- Provide quantitative expressions and data for the developed qualitative description of the problem in order to provide a defined solution.

- Perform sensitivity analysis of the assumptions and data to check the robustness of the solution.

In his book about environmental negotiations, Gorczynski (1992) suggests several steps through which a manager can organize to prepare:

- What are we taking action about and why?

- When will we take action and where?

- Who is involved and how are they arrayed?

Eisenhardt, *et al.* (1997) suggests six tactics for managing conflict:

- work with more, not less, information;

- develop multiple alternatives,

- share commonly agreed-upon goals;

- interject humor into the decision process;

- maintain a balanced power structure; and

- resolve issues without forcing consensus.

Building on the discussion of environmental risk management in Chapter 2 and interfacing the assessment, management, and communication of risk in Chapter 3, and applying the principles of the authors cited above, we now describe the practice of Risk-Based Analysis (Table 4.1). We believe that this five-step procedure will be readily adaptable to most corporate processes, and can serve as an organizing theme for mapping a strategy for an impaired property. The aim here is provision of a practical technique that begins "with the end in mind," and using a systems approach to integrate good science and business practice, with good communications to positively effect the environmental risk management process and thereby lead to a more sustainable urban-industrial landscape. To help, we developed Appendix D using a workbook-like framework to provide an outline and checklist of this procedure in tabular/graphical form.

Table 4.1
Risk-Based Analysis:
Five Progressive, Knowledge-Building Value Points

Problem Formulation

First, define the problem; specify needed resources, deadlines, and scope. Use conceptual models to guide definition of source (cause), effect, and the many influencing factors. Establish the boundaries and operational context of the problem and the associated impairment or risk issue(s). Develop a preliminary model of the decision-making process and identify data needs to inform that process and define the necessary quality of data (*i.e.*, if you collect or calculate it, will it convince).

Situational Analysis

Identify, understand, and integrate the needs and objectives of others within the regulatory, political, and socioeconomic aspects of the property and their roles in risk management decision-making.

Risk Assessment

Quantify and qualify the nature, frequency, and intensity of risk. Set the scientific data and findings in redevelopment/reuse contexts.

Risk Management Option Development

Depending on the problem and its situational context, address what options are available to scientifically and justifiably explain away the reputed risk or impairment, cut-off exposure pathways (and therefore risk), or permanently reconstruct the "risk system" (*i.e.*, source or effect) so that it no longer exists, or is quantitatively reduced in magnitude by a significant and sufficient degree. In addition, to help influence outcome options, develop your risk mitigation scenarios within the context of redevelopment and economic revitalization. It may even be worthwhile to develop a short- and long-term amortization of risk over a sufficiently long planning horizon to better contain costs and land use.

Risk "Argument"

In this step, develop a convincing communications approach to achieve optimal, "mutual gain" solutions by integrating property value, environmental risk or impairment, the situational context, risk-management options, and decision-making frameworks. The risk information and preferred risk management option are formulated within a communications program by which it is presented and, ultimately, negotiated into an approach acceptable to all.

II.   PROBLEM FORMULATION

A.   DEFINE: PROBLEM, DECISION, DATA, AND PROCESS

The *presenting problem* is the triggering event or issue that brings an impaired property matter to the fore (Smith, 1998). While it is not necessarily the specific thing that is wrong or in need of improvement, it does provide the starting point for identifying and defining the problem(s), their associated decisions, necessary data/ information, and the process. The presenting problem is often one or a combination of the agents of change spoken of at the very start of Chapter 1.

Problem formulation is performed as a process, yielding a useful product (Smith, 1998).

- As process,

- o the manager may identify and validate what the problem is, and provide boundaries and a conceptual framework of what the problem is and is not,
- o characterize various elements within the problem and their interrelationships, and
- o finally, develop an interpretation of the whole situation in the light of relevant knowledge.
- As product, the manager documents their concept of the matter, and presents a problem-solving process plan to guide follow-on steps.

We see problem formulation as a four-phase protocol (Figure 4.1) similar to Smith's (1998) eight-step method called *Situation Definition and Analysis* (or SDA). The relationship between our problem formulation phases (primary bullets) and Smith's SDA method (secondary bullets) is shown below.

- Define the problem
  - o state the presenting problem
  - o analyze the presenting problem
  - o broaden and deepen the analysis
  - o bound the problem
  - o diagram the problem
- Define the decision(s)
- Determine how to inform the decision
  - o identify topics and issues requiring more study
- Develop a management plan
  - o develop an overall problem-solving plan
  - o develop a summary account of the situation

Problem formulation is not static but an ongoing process that can ensure that information and understanding evolve, so that the assessment efforts, management strategy and tactics, and communications are focused on the most important issues and needs of those within your company or unit and without, especially stakeholders.

B. DEFINE THE PROBLEM

Defining the problem provides the initial objective description of the system or situation to be evaluated. It involves stating the presenting problem, generating and evaluating preliminary hypotheses about causes (that is, sources of impairment such as chemicals, hypothesizing about primary and secondary mechanisms controlling their release and movement in the environment, hypothesizing about possible environmental transport pathways), and identifying possible current or reasonably anticipatable human health or ecological targets as well as important community issues. Conceptual models are subsequently prepared to describe the system/situation under consideration.

Defining the problem also helps the manager develop management goals and endpoints for the risk assessment itself; in this way the manager begins to manage the definition of the human and ecological risk—the defined peril—as a result of the process outlined in Figure 2.1:

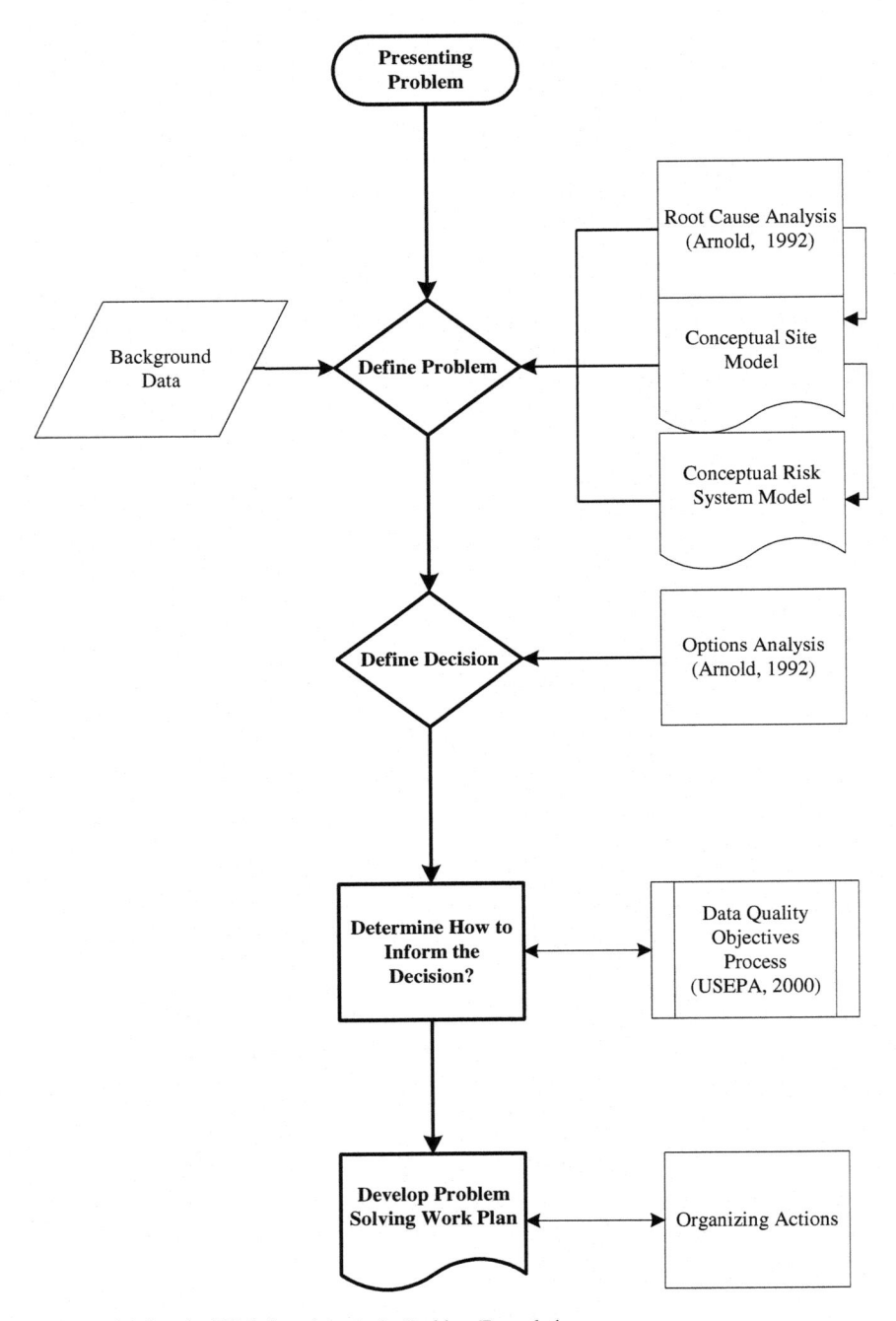

Figure 4.1 Step 1 of Risk-Based Analysis: Problem Formulation.

- Management goals—are not just for the company (in terms of legal, image/relationship, and financial, as described in Chapter 2 and Appendix B), but those situations, issues, or concerns that require resolution at the impaired property, as well as the realization of opportunities through the re-designing of the risk system. Their selection depends on the type of hazards actually presented or supposed to be present (definition of risk as harm, its importance, and significance), the human, ecological, or cultural resources at potential risk, what stakeholders are involved, various constraints (which might be present or develop), and regulatory requirements.

- Assessment endpoints—these endpoints interpret the management goals for the system/situation of concern:
  - They are explicit expressions of the actual human, environmental, and cultural values one is interested in protecting. Two elements are required to define such endpoints.
    - identification of the specific valued entity (*e.g.*, human population, species, functional groups of species, a community, etc.)
    - identification of the characteristic of the entity of concern that is important to protect and is potentially at risk (*e.g.*, public health, nesting, nutrient cycling, and integrity of an important cultural circumstance)

    Assessment endpoints must be relevant to the situation and be susceptible to the impairment (cause and effect).
  - Often, arriving at assessment endpoints is an interactive process that incorporates stakeholder views and technical feasibility to refine the goals. Nevertheless, trying to define them as early as possible is key.

- Measurement endpoints—these endpoints provide tangible techniques for measuring characteristics relevant to the assessment endpoint. For example, if the assessment endpoint is the reproductive success of robins, the measurement endpoint could be egg production or fledgling success.

- Conceptual models describe the presenting problem and key issues and relationships—they include written descriptions and visual representations of the situation and risk system (Figure 1.1 and 1.2). They form the basis for determining what data are available or needed and what analyses will best evaluate the risk hypotheses, the decision-making that is likely to be operating, and techniques for controlling or eliminating the impairment.

The sections below describe the problem definition protocol.

1. Assemble Available Background Information

First, collect all available information. Following standardized procedures will be helpful:

- Standard Guide for Phase I Environmental Site Assessments E1527 (ASTM, 2000)

- Standard Guide for Property Condition Assessments E2018 (ASTM, 1999)

Additionally, the Risk Appraisal approach (Frantzen, 2000) may prove useful in collecting baseline information. Table D.2 in Appendix D provides a checklist of likely background information, such as data and maps, which should be available during the problem formulation stage.

2. Analyze the Problem

In order to define the problem objectively it is necessary to analyze the available information. One way to accomplish this analysis is with Arnold's (1992) Root Cause Analysis process, which focuses on getting to the "why" of the problem. Arnold's approach is similar to the Kepner-Tregoe (1981) method of *Problem Diagnosis*, which applies a comparative technique between the defined problem and similar situations. It also is similar to the *5W2H* method of Robinson (1993) (*i.e.,* what, why, where, when, who, how, and how much) as will be seen. The following subsections describe the process and Table D.3 provides a summary outline.

a. Prioritize

Using the available information, list and prioritize obvious issues and concern according to their likely impact, urgency, amount of data, ability to influence, severity, probability, and/or potential (financial) loss. This step is similar to Grose's (1987) "Hazard Totem Pole" approach, as discussed in §III.A of Chapter 2. This step helps focus and order the following effort.

b. Define/Describe

Using general, observable, factual information, describe the boundaries of the presenting (environmental) problem in terms of identity (who/what), location (where), and timing (when). As necessary, describe not only what is happening but also what is not happening. Precision is important here. Furthermore, prepare a statement describing how you know about the problem and how, if necessary, one could verify it further.

c. Distinguish

Next, state what is distinctive about each of the boundaries of the presenting (environmental) problem in terms of characteristics, scope, and/or magnitude: identity—who/what; location—where; and timing—when.

d. Diagnose

At this stage, using the available descriptions and distinctions it is necessary to draw preliminary hypotheses as to what may be the source(s) of the presenting (environmental) problem(s) at an impaired property. It is generally beneficial at this stage to graphically present the diagnostic explanation as a conceptual model, which is described in the following section.

e. Test

Next, test (mentally or using available data) the developed diagnostic hypotheses for reasonableness and the ability to subject them to further verification (measurement).

f.   Decide

Finally, using the preliminary hypotheses and the conceptual models, draft a preliminary set of possible decisions concerning the definition of the presenting (environmental) problem(s) and likely actions:

- management,

- assessment,

- corrective (remedial),

- preventative, and

- opportunistic (re-designing the risk system and coordinating it with redevelopment/reuse possibilities).

3.   Conceptual Modeling to Aid Problem Definition

a.   Approach

Conceptual models describe, qualitatively and/or quantitatively, relationships (including possible causal relationships) among the abiotic systems (physical system drivers and chemical stressors), biotic systems (human and ecological), and cultural systems (including land use and landscape activities), as suggested in Figure 1.1. In essence, such models are a formal statement of causal pathways for human and ecological risk (the defined peril) that leads to the development of a suite of testable hypotheses and possible actions for managing the full spectrum of potential environmental risks. If possible, depending on available data, rank these risks and use them to prioritize activities and the allocation of resources. The ultimate quantification of the elements comprising the conceptual models provides the scientific basis for the quantitative estimation of risks.

The purpose of these models is to provide the stakeholders and decision-makers with (Suter, 1999 and USEPA, 2000):

- a basis for thinking through the environmental risk system and framing various worldviews,

- a clear spatial and temporal understanding of the site (that is the property and its surroundings),

- aid in the identification of actual or potential sources and potential environmental transportation pathways,

- an understanding of likely sensitive human and ecological receptors or targets that might be exposed and their spatial/temporal relation to the site,

- insight into possible complete routes of exposure,

- clarification of assumptions concerning the situation to be assessed,

- a communication tool for conveying assumptions to risk managers, and

- a basis for organizing and managing various follow-on tasks, including the risk assessment.

b.   Types

Concept mapping is a general method that can be used to help any individual or group to describe their ideas about some topic in a pictorial form (Smith, 1998). It is a structured process focused on a topic or construct of interest that produces an interpretable pictorial view (concept map) of their ideas and concepts and how these are interrelated. In the traditional EPA risk assessment paradigm, conceptual models represent the release of a contaminant, its transport through the environment, and its contact with a receptor (USEPA, 1989b, 1998b, and 2000). A flow chart generally represents these models (Figure 1.2).

Suter (1999) suggests that these models:

- be explicitly mechanistic,

- define all important functional compartments and components that define the risk system to be managed,

- show exposure and response relationships,

- be split into multiple hierarchical models to enhance understanding, and

- present modular components or issues that may be observed, measured, assessed, managed, and/or controlled.

Variations of the concept map idea are useful for detailing the system under assessment. In Risk-Based Analysis, we believe that the two models of greatest value are the Conceptual Site Model (CSM) and the Conceptual Risk System (CRSM) Model. Together these models provide a picture or framework of the manager's problem. It sets forth the tasks facing those preparing the environmental and especially the risk assessment. It also depicts those issues that concern the decision-makers and the public. Any model should be "living," in that it is regularly updated as additional data are collected, knowledge gained, and knowledge frames (gestalts) evolve.

c.   Conceptual Site Model

A CSM graphically describes the site, its physical and cultural environs, and ecological features (Figure 4.2 and Table D.4). It also provides indications of modes of transport, *i.e.*, typical wind direction and speed, as well as directions of overland flow of water and drainage, surface water flow, and groundwater flows. This information, along with the identification of potential sources of hazardous materials, provides insight to the potential transport and fate of released chemicals into the environment. Finally, it can provide not only spatial information but also temporal information in the form of historical site layouts, seasonal activities, important cultural items, and proposed or likely land use changes/developments.

d.   Conceptual Risk System Model

The CRSM picks up at this point and provides a graphical depiction of the spatial and temporal relationships between hazardous material sources and their fate in the environment to a potentially exposed human population, biota, critical habitat, or cultural area/fabric (Figure 4.3 and Figure D.1).

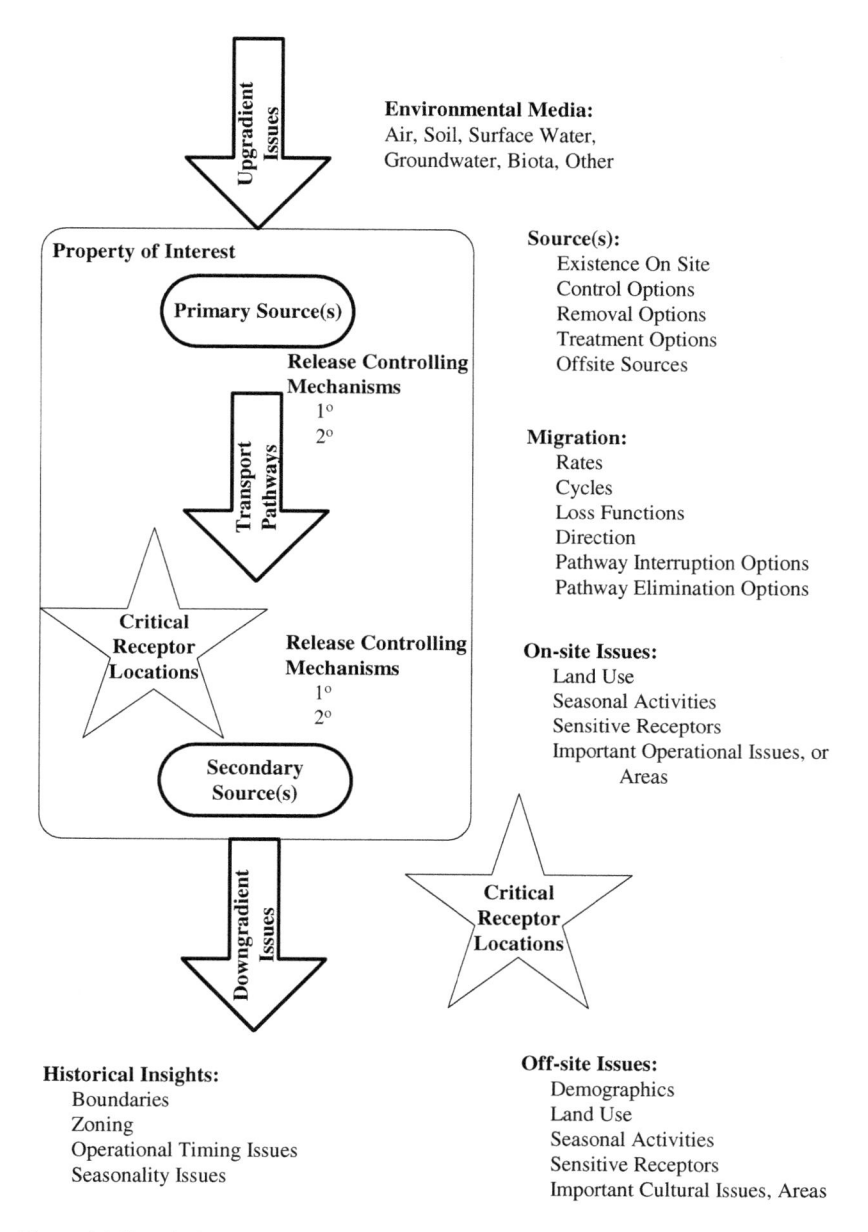

Figure 4.2 Generic Conceptual Site Model.

The CRSM also illustrates exposure scenarios. These scenarios incorporate information on exposure pathways, exposure points, and potential receptors; and they provide hypotheses about what may be happening at the site now or in the reasonably anticipatable future.

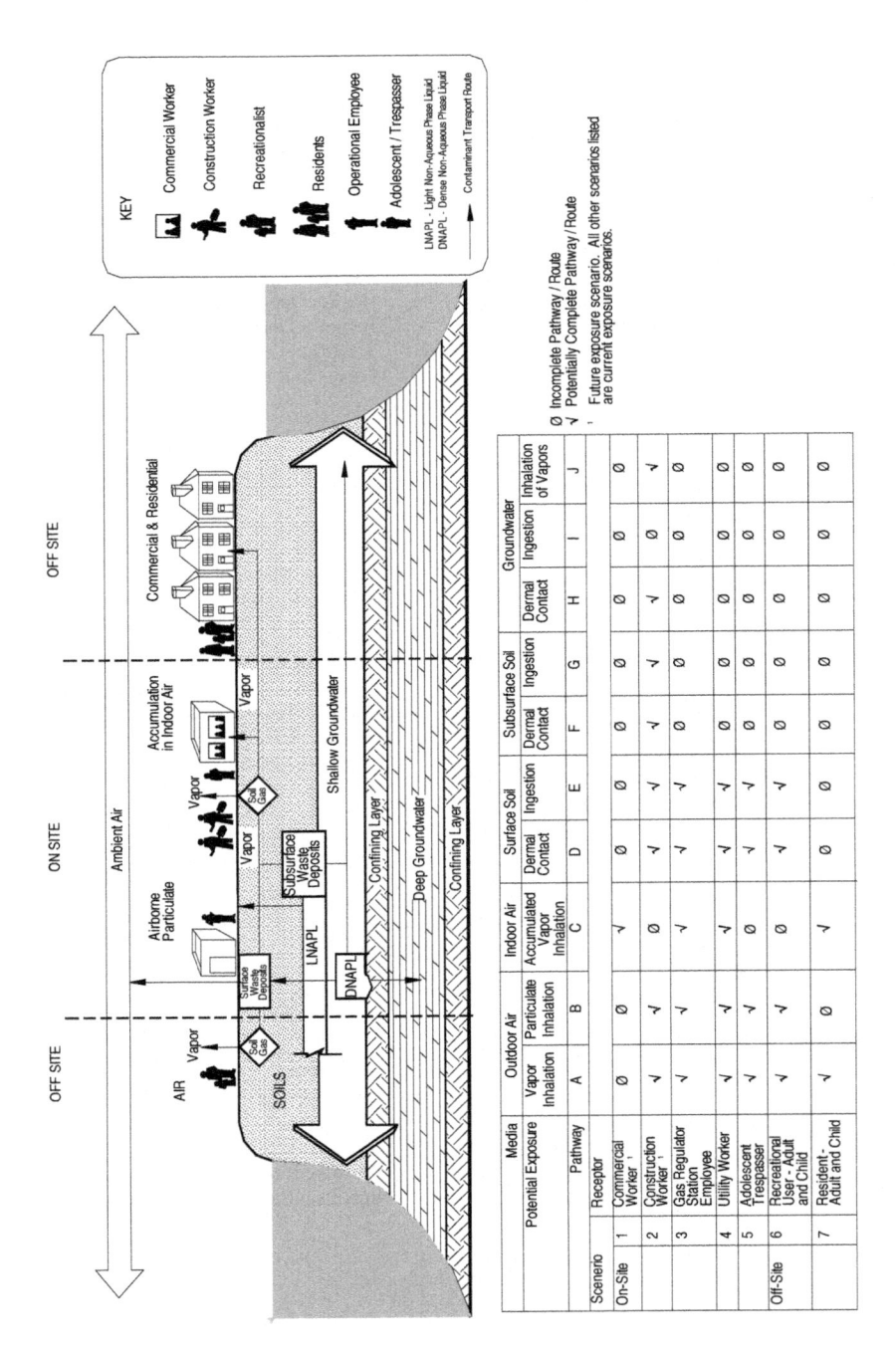

Figure 4.3 Conceptual Risk System Model.

C. DEFINE THE DECISION

The next step in problem formulation is the definition of the decision or decisions that might be required (Table 4.2). At this stage, it is necessary to identify as many of the objective issues associated with the environmental problem as possible and clarify why a decision, or even a solution for that matter, is necessary (see Table D.5 and D.6). This illumination of issues and their prioritization should reveal some criteria, parameters, or specifications important to decision-making. This information leads to the development of a decision statement associated with the previously defined problem (Table D.7). This formal statement defines what needs to be determined, in non-binary terms, and why. The statement should identify what one is trying to achieve, preserve, and avoid as tangential problems by whatever is decided. It also is possible to begin identifying, testing, and troubleshooting possible risk management options. This is important at this stage because possible solutions require the collection and consideration of certain information so that one may select or reject them (Smith, 1998).

D. INFORM THE DECISION: DATA QUALITY OBJECTIVES

After defining the problem and decision, one needs to determine what data need to be collected to inform the decision-making process and lead to resolution of the environmental issues. Perhaps the best way to accomplish this phase of problem formulation is with the Data Quality Objectives (or DQO) Process. The DQO process is USEPA's strategic planning approach, based on the scientific method, to prepare for a data collection activity (USEPA, 2000). It provides a basis for balancing decision uncertainty with available resources, time, and money. It also provides a systematic procedure for defining the criteria that a data collection approach should satisfy, including when to collect samples, where to collect samples, the tolerable level of decision error for the study, and how many samples to collect, while balancing risk and cost in a disclosable manner. The process has two major activities:

- specifically state the question(s) that need to be answered for the problem and decisions at hand (note relationship to early problem formulation steps), and

- specifically state the amount of uncertainty you, the process, or the stakeholders are willing to tolerate in the data and when answering the questions at hand.

This process is a planning tool that can save resources by making data collection operations more resource-effective. Tables 4.3 and D.8 summarize the seven-step process and indicate relationships to other problem formulation steps.

E. PLAN THE REST OF THE PROCESS

Problem formulation "sets the stage" for the remaining steps of Risk-Based Analysis. According to Smith (1998), complex problems, such as environmental risk management issues, are protracted efforts requiring management. The problem solving plan step of SDA is useful here in that it identifies "where things are going and how are we going to get there." Following Ostrow's suggestions in Appendix C

Table 4.2

Problem Formulation: Defining the Decision

Element	Description
Illuminate Issues	The corporate manager should eliminate the non-decisions early in the process. Ask: has something happened that should not have, or is something missing that needs to be provided. If the answer is no, then perhaps no decision is necessary.
State Decision/Decisions And the Purpose	The decision purpose needs to fit into the larger picture of environmental liability. Broad statements are better here to allow for the widest set of solutions that can be tailored to corporate needs.
Develop Preliminary Decision Criteria	These criteria should state what you want to achieve, preserve and avoid through the decision making process. Positive criteria are best, try to eliminate "avoids" and seek "achieves." This will focus the decision process and reduce redundancies.
Prioritize	Certain criteria or issues should now stand out. These issues lead to the formation of decision criteria. From this list, differentiate between what is essential (absolute requirements) and what the corporation ideally would like to achieve (desirable objectives). Absolute requirements may revolve around costs, performance levels, regulatory criteria, etc. Absolute requirements focus the decision-making process. Eliminate options that do not meet the requirements. Rank desirable objectives.
Identify Preliminary Risk Management Options	List all the ways/options you might use to meet the decision criteria.
Early Evaluation of Options	Next, test options against the criteria. Complex decisions with multiple absolute requirements and desirable options can be breadboarded in a matrix. Decide for each option whether it satisfies every absolute requirement. If an option fails this, disqualify it and move on. Once judgments are made based on the absolute requirements, move on to the desirable objectives. Options with highest score are best.
Troubleshoot and Refine	How can the best option be further refined and what could go wrong with the communication and implementation of it?

Based on Arnold, 1992.

Table 4.3
Data Quality Objective Process

Step	Purpose	Activities	Outputs
Identify Problem	Clearly define problem requiring new environmental data to focus the study clearly and unambiguously	Concisely state problem. Establish Planning Team & Identify decision maker(s) and specify available resources and relevant deadlines	Concise problem description, list of planning team and decisionmaker(s), available resource summary & schedule
Identify Decision	Define the decision that will be resolved using data to address the problem	State decision(s); if multiple, categorize them. Also, state potentially resulting actions or outcomes.	Decision Statement and Actions/Outcome List
Identify Decision Inputs	What information is necessary to make the decision? Which inputs require measurements?	Identify needed information and sources, what is needed to establish an action level or decision criterion, and confirm techniques and methods	List assumptions made and inputs, variables, and characteristics to be measured.
Define Study Boundaries	Specify the spatial and temporal circumstances covered by the decision	Define geographic area covered by decision(s). Specify the characteristics that define things of interest. When appropriate, divide the things or area of interest into "strata" with relatively homogeneous characteristics. Define time scale of decision and data collection timeframe. Identify constraints on data collection.	State characteristics defining the problem domain. Detail the spatial/ temporal boundaries of the decision. List practical constraints
Develop Decision Rule	Integrate the outputs from previous steps into a single statement that describes the logical basis for choosing among alternative actions.	Specify the parameter that characterizes the population, area, or things of interest and an action level. Combine the outputs of the previous steps into an "if...then..." decision rule that defines the conditions that would cause the decision maker to choose among alternative actions.	An "if...then..." statement that defines the conditions that would cause the decision maker to choose among alternative courses of action.
Specify Decision Error Limits	Specify the decision maker's acceptable limits on decision errors, which are used to establish appropriate performance goals for limiting uncertainty in the data.	Determine range of the parameters of interest. Define types of decision errors and identify potential consequences of each. Specify a range of possible parameter values where the consequences of decision errors are relatively minor (gray region). Assign probability values to points above and below the action level reflecting the acceptable range of decision errors. Check the limits on decision errors to ensure that they accurately reflect the decision maker's concern about the relative consequences for each type of decision error.	The decision maker's acceptable decision error rates based on a consideration of the consequences of making an incorrect decision.

Table 4.3

Data Quality Objective Process

Step	Purpose	Activities	Outputs
Optimize Data Collection Design	Identify the most resource-effective sampling and analysis design for generating data that are expected to satisfy the DQOs.	Review DQO outputs and existing data and translate the information into statistical hypothesis. Develop approach alternatives, and for each alternative, formulate approaches for solving design problems. Also, for each alternative, select optimal sample size that satisfies the DQOs. Select the most resource-effective design that satisfies all objectives, and document operational details and assumptions of the design in a Sampling and Analysis Plan.	The most resource-effective design for the study that is expected to achieve the DQOs, selected from a group of alternative designs generated during this step.

After USEPA, 2000.

we see the fourth phase of problem formulation resulting in a management plan for the process. Although focused on risk communication and the development and implementation of a strategy and plan, Ostrow discusses four crucial priorities in Appendix C:

- define the risk,

- identify the cause of the risk,

- describe actions that mitigate the risk, and

- demonstrate responsible management action.

Rosengard in Appendix B identifies the following crucial issues in dealing with the financial implications of environmental risk:

- Managers must avoid being out-negotiated by contractors, regulators, insurers, and other potentially responsible parties.

- Managers must avoid inconsistent strategies and costs for similar projects.

- Managers must understand how spending narrows cost ranges and reduces liabilities.

- Company decisions must consider lifecycle costs, reimbursements, capital expenditures, operating business impacts, land use issues, property taxes, and property value from decision analysis.

- Mere project completion is not the management endpoint; the endpoint is the satisfaction of stakeholders and regulators and the limiting of exposure to current and future business risk.

It also is important to begin to consider and if possible quantify within a reasonable order of magnitude the physical, market, and environmental regulatory conditions that will influence the feasibility of redeveloping sites to their highest and best potential (Ackerman *et al.*, 1998). As suggested in Figure 3.7 by the interaction of planning and Risk-Based Analysis, preliminary planning will stimulate the entire process through the consideration of redevelopment option for its risk potential and cost effectiveness in relationship to remedial action alternatives.

Taking these points to heart early in the process and integrating them with the technical requirements of the work (remember the DQO process) are essential in the development of the objective aspects of the management plan for proceeding (Figure D.9). This planning step focuses on the objective aspects of the risk problem. In the situation analysis stage of Risk-Based Analysis, the manager engages in deducing who the players are in the process and what their issues are or are likely to be. Based on this information, the manager may consider reformulating the problem and decision descriptions and develop a plan for the establishment and management of an extended knowledge-based network that includes those external to the company.

III.   SITUATION ANALYSIS

A.   INTEGRATING PERSPECTIVES, ANTICIPATING ISSUES

We defined Problem Formulation as a process to identify, characterize, bound, and validate the problem, and document a manager's concept of the matter together with a problem-solving process plan to guide follow-on steps. Its primary focus is on the objective, technical, legal, and financial aspects of the problem, and the involved decisions. In contrast, *Situation Analysis* seeks to help the manager identify, understand, and integrate the needs and objectives of others within the regulatory, political, and socioeconomic aspects of the environmentally impaired property problem and their roles in risk management decision-making (Figure 3.7). We follow the suggestion of Ostrow (see §V in Appendix C) urging the use of situation analysis as an important management and communications planning tool; and we see this analysis as a way of identifying others outside of the company and developing an appreciation of their issues (Figure 4.4). Subsequently, through the development of a communications plan the manager can begin to plan how to engage others outside of the company, develop explanations of the problem and the overall process, and determine how to ensure expectation clarity as well as incorporate stakeholder values in the process.

B.   PLAYERS

As already discussed (§V.A of Chapter 3), environmental issues involve a great number of parties within an organization (as discussed above) and without (Figure 4.5 and Table D.10). Stakeholders see themselves as affected (in one way or another) by a particular environmental issue or problem; they may have information, knowledge, resources, or positions that are important in relation to the issue or problem; and, they may control mechanisms instrumental for intervening. Therefore, identification of the appropriate stakeholders for inclusion is critical early in the process. We believe it important to create a situation-specific network through which relevant groups (*i.e.*, community, regulators, etc.) can interact. It is necessary to consider the local power structure (both in the informal and formal sense) and those in closest proximity of the site. The outcome of this identification process (Figure 2.3) is the ability to better inform and refine the problem statement and decision process plan previously developed during the problem formulation stage.

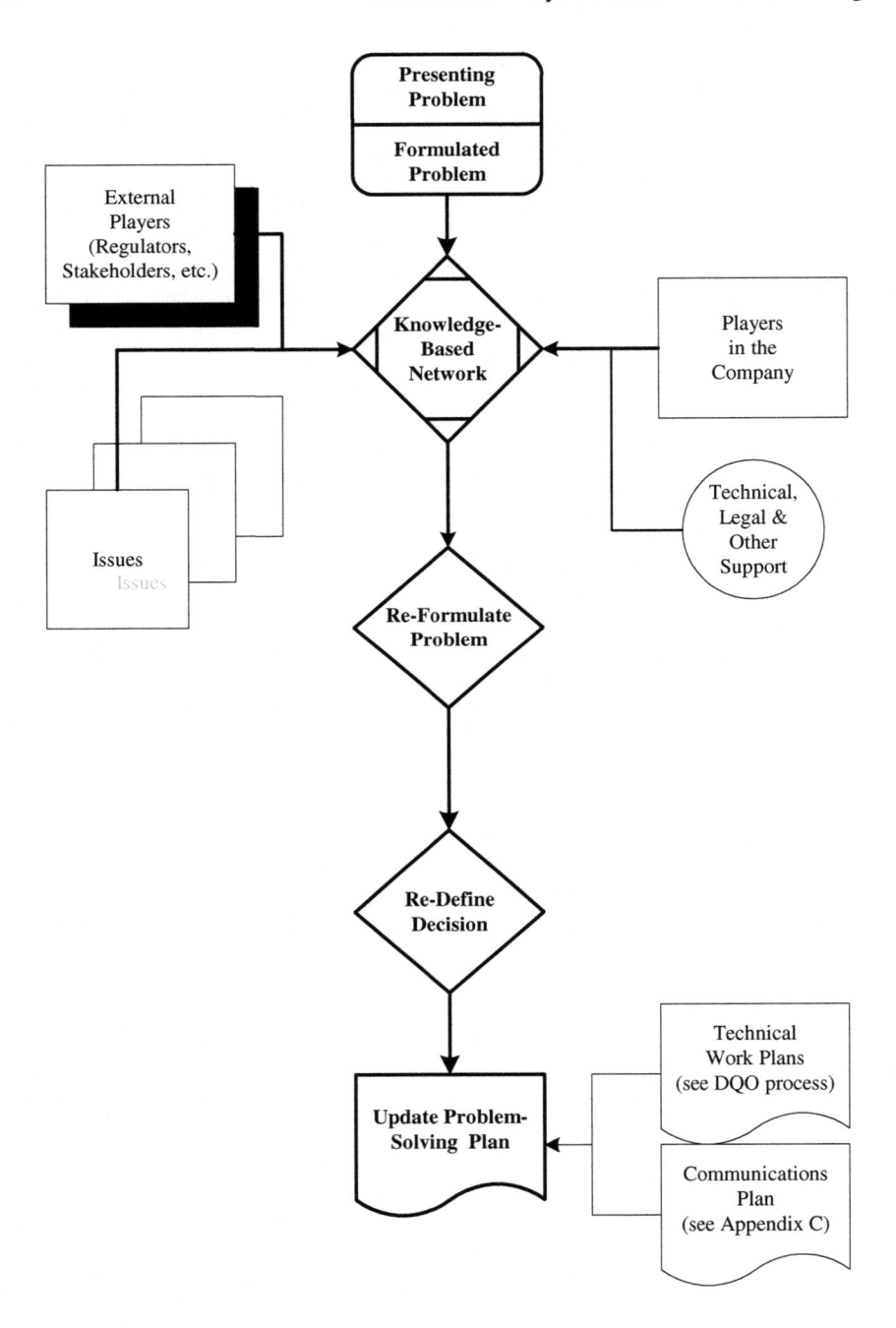

Figure 4.4 Step 2 of Risk-Based Analysis: Situation Analysis.

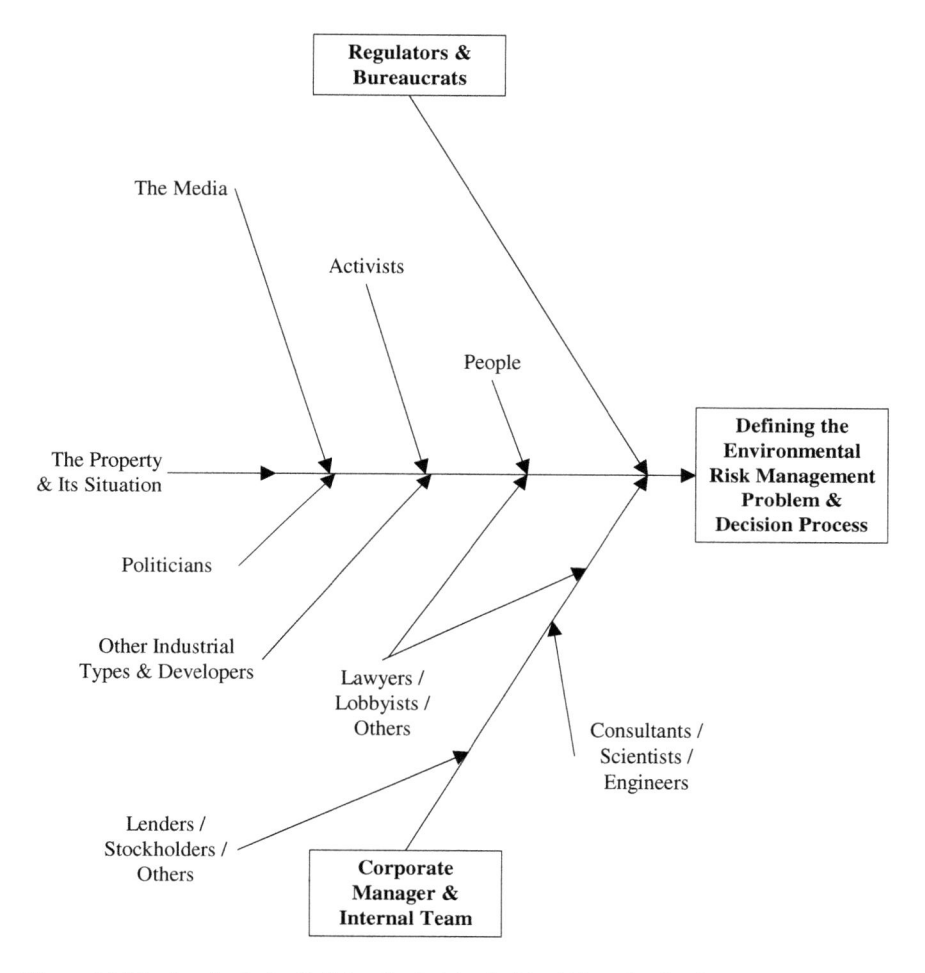

Figure 4.5 Situation Analysis. Defining the decision by identifying who the players are, as suggested by Gorczynski (1992).

C. ISSUES

Now that the fundamental problem and the various players involved in the decision process are identified, it is important to associate the many players with the issues associated with the problem and the decision. This involves a re-evaluation of the scope and priority of previously identified issues and identification of subjective issues crucial to managing the definition of risk (Tables D.11 and D.12). This re-evaluation should incorporate, as suggested by Llewellyn (1998), the critical concepts of (see the discussion in §V.B of Chapter 2):

- how the various players interpret what is of importance in terms of risk,

- how risk is defined as harm,

- how the significance of risk should be evaluated, and

- concerns about uncertainty.

D. PROBLEM FORMULATION UPDATE

As suggested by Gorczynski (1992) and Robinson (2000), and recognized in Figure 2.7, integrate the objective and subjective aspects of the problem to better define the decision context and process. This helps the manager anticipate pressure that will come to bear due to the interplay of real and perceived risks, and the ethics and values of stakeholders (see §IV.B of Chapter 2 and Robinson, 2000). While preparing, gather and consider information and guidance from the financial/risk management, operations, and planning functions in the company in the situation analysis (Figure 3.7).   Then, use Strengths-Weaknesses-Opportunities-Threats (SWOT) Analysis to reformulate both the problem and decision statements (Figure D.13). Next, update the management plan developed previously with what you have learned. Finally, after reformulating the problem and updating the management and communication plans with realistic goals based upon an understanding of the external knowledge-based network one is interacting with and the issues of the associated players discharge the following suggestions of Gorczynski (1992):

- Educate and mobilize the players on the team.

- Actively approach potential allies; if they share a common interest with you, persuade them to join your team.

- Educate and mobilize allies successfully recruited.

- Delegate and take action.

IV.   MANAGING THE RISK ASSESSMENT

Risk assessment is an objective measure of a subjective subject requiring the balancing of what management wants with what science can do. The environmental risk system classically has been seen as consisting of the following (remember Figure 1.2):

- Source (primary)

- Release Point

- Transport

- Fate, possibly resulting in a Secondary Source

- Release Point from Secondary Source

- Transport from Secondary Source

- Exposure Pathway

- Exposure Point

- Exposure Route

- Receptor (*e.g.*, humans, biota, habitat, cultural resource areas)

In reality, the risk system begins along this exposure corridor and then radiates outward. Beyond the potential health and ecological risk, it will cause cleanup or remedial liability and compliance liability; it also may cause adverse impact upon the value of real property (at and near the proximity of the site), socio-cultural and socioeconomic impacts, as well as the potential for third-party liability and litigation.

Environmental decision-making often seems driven by a belief in the safety of numbers but defining what is acceptable complicates the process. This is most evident in the contrast of scientific estimates of risk (assumed as reasonably accurate reflections of reality) and others' risk evaluations (which will reflect differing gestalts). These different modes of evaluation often conflict and tension is inevitable. Furthermore, regulatory decisions based "on policy" create more tension between the company and regulators as well as the regulators and "those at risk."

Scientific knowledge feeds risk assessment (Figure 2.1). An assessment results in a characterization of the potential adverse health and environmental risks associated with the impaired property. This result informs the risk management function leading to decision-making. To prevent problems, environmental decision-making needs centering within an understandable, common-sense context that considers science, finance, legal issues, and that values stakeholders. Risk assessments are criticized for focusing on only a narrow set of outcomes, *e.g.*, certain human health hazards. Thus, risk characterizations should address social, economic, ecological, and ethical outcomes as well as consequences for human health and safety (NRC, 1996; see Appendix A). Risk characterizations should also address outcomes for particular populations in addition to risks to whole populations, maximally exposed individuals, or other standard affected groups. Adequate risk characterization depends on incorporating the perspectives and knowledge of the spectrum of interested and affected parties from the earliest phases of the effort. Further, the breadth of analysis and the appropriate extent of involvement or representation required for satisfactory risk characterization are situation-dependent. Achieving this context requires that the assessment of risk include both quantitative and qualitative risk arguments about actual risk. As suggested by Figure 3.1, the risk assessment also must include, or be contextualized by, an understanding of perceived risk (developed during situation analysis as well as by ongoing risk communication).

Moreover, the design of environmental assessment reports (including risk assessments) is to be National Environmental Policy Act (NEPA) compliant or equivalent, and so address the public and stakeholder value concerns of the "commons." This includes the site within its community context. In this way of thinking, a risk assessment is not an academic exercise or science experiment. It is applied analysis that translates science into useful information for management and policy decision-making. Table D.14 outlines the typical needs of a baseline risk assessment as defined by USEPA (1997g). However, as we have discussed, the assessment report must provide a risk characterization responsive to the issues and concerns identified as important and significant by stakeholders and not just regulators (see discussion in §V.B below and Figure D.15).

The manager should look to develop a multidisciplinary team of focused technical experts with in-depth knowledge of different fields and the relevant regulations and guidance, particularly in toxicology, public health, ecology, and various biological sciences as well as planning, socioeconomics, and other allied fields. The team also should include appropriate company representatives, legal counsel, and others as necessary to contribute to the formulation, implementation, and review of the assessment (see Figure D.15). Whoever will be managing the assessment for the company should be a seasoned professional with a trans-disciplinary understanding of environmental risk problems, able to integrate people and findings by applying creative strategies, while basing the program on sound science.

Determining the manager's needs and expectations is a key step in designing the assessment. Murphy and Fitzgerald (1994) recommend developing a formal written assessment plan that clearly states its purpose, objectives, scope, schedule, and methods to help the risk assessor and manager to:

- clarify the manager's needs and expectations;

- assure that the manager gets the analysis they expect; and

- assure that the manager clearly understands the results and appreciates the uses and limitations of risk assessment. Such explanation is critical as it sets the stage of expectations.

Basic questions arise concerning methodologies or the significance of the expected results at many points during the assessment. Consequently, risk assessors should meet periodically with the manager to assure maintenance of direction, understanding, and approval. The assessor or team should expect requests for refinements in the objectives of the risk assessment as it progresses. Finally, recognize that risk assessments have significantly reduced impact if their results are poorly documented or communicated. The "numbers" are many times the least useful, yet they typically occupy the most prominent attention in the assessment and in people's minds, at the expense of descriptive findings and interpretative dialog.

The integrated project team can and should provide input during the problem formulation and even situation analysis stages of the company's effort. This should then translate into the design of the assessment's approach, context, style, and focus. The design should consider the decision-making process, who is making and influencing the decision, and what information they need. We think it important to involve a risk assessor or team members experienced with risk assessment early in the process to help the entire corporate team appreciate what the science can and cannot support, and what the applicable regulations seek and risk assessment guidance allows. At the same time, if the risk assessor does not begin with the end in mind, corporate needs, and those needs of regulators and public stakeholders, will more than likely not be met.

V.   RISK MANAGEMENT OPTIONS

A.   BASIS FOR ACTION

The decision of what is or is not acceptable is grounded in the particular regulatory framework under which the site is assigned, agency acceptability and community/stakeholder acceptability as they are expressed in the importance, harm, significance, and uncertainty of risk (see §V.B of Chapter 2). Table D.16 outlines

USEPA's basis for action, and many states follow a similar decision basis, although the triggering threshold may differ. If acceptable, no further action is generally indicated. However, if judged unacceptable, then remedial action is necessary to decide what to do to correct the situation, and determine how to accomplish such action.

B. GOALS, OBJECTIVES, AND ENDPOINTS

Developing remedial goals is easy; one just runs the typical risk assessment in the reverse direction. Getting others to agree is another matter. Classically, EPA operates with a "Bottom-Up" approach beginning for example with a chemical concentration equal to a cancer risk, for example, at 10^{-6} and going up from there if they believe that it is justifiable on a variety of objective and subjective (policy) grounds. They prefer the lowest risk possible and permanent removal or treatment of sources of risk. Many state regulatory agencies adopt a simple "Middle-of-the-Road" approach electing to base all decisions on a threshold in the middle of the classical risk range. These approaches contrast to the last two polar opposites: the "Top-Down" approach used by many responsible parties that is based on offering remedial goals based on the maximum acceptable target risk levels and being negotiated down to a lower level; and the approach used by many NGOs and often the public, that is, wholesale removal of every toxic molecule.

The issue of "how clean is clean" is resolved by designing risk management options around an argument about risk reduction based on eliminating sources of risk, preventing exposure, and/or monitoring (Figure 4.6). The argument must address both objective aspects and subjective aspects. The objective issues include who or what are you trying to protect, the need to meet regulatory requirements, a range of possible numerical concentration targets based on the acceptable range or "threshold" level set by the regulatory agency, and the techniques to be used to temporarily or permanently reduce risk or limit exposure. There may be benefit in some cases to using past regulatory cleanup goals, scientific analysis of background levels, volumetric considerations, and/or a description of how a particular chemical or mixture will behave in the environment to support the argument (LaGoy and Hopkins, 1991). The subjective issues include addressing people's concerns about safety, demonstrating how the remedial techniques will ensure risk reduction and/or limit exposure, and illustrating how the action will contribute positively to the community. As Carpenter (1995) suggests, risk assessors need to explain, in lay terms: what we know, what we do not know or are unsure of and why, what else could we know, and what is it that we should know to act with these uncertainties?

Periodic monitoring of the remedy (especially when Institutional Control measures are used) assures that it is operating as per its design, to reduce the concentration of chemicals in a timely, cost-effective manner (see the end of Figure 2.10). A number of strategies are useful in improving performance and shorten cleanup timeframes:

- Implementing remedies in multiple phases—may increase the performance and cost-effectiveness of the long-term remedy. Performance data from an early phase can be used to refine the design of later phases so that the

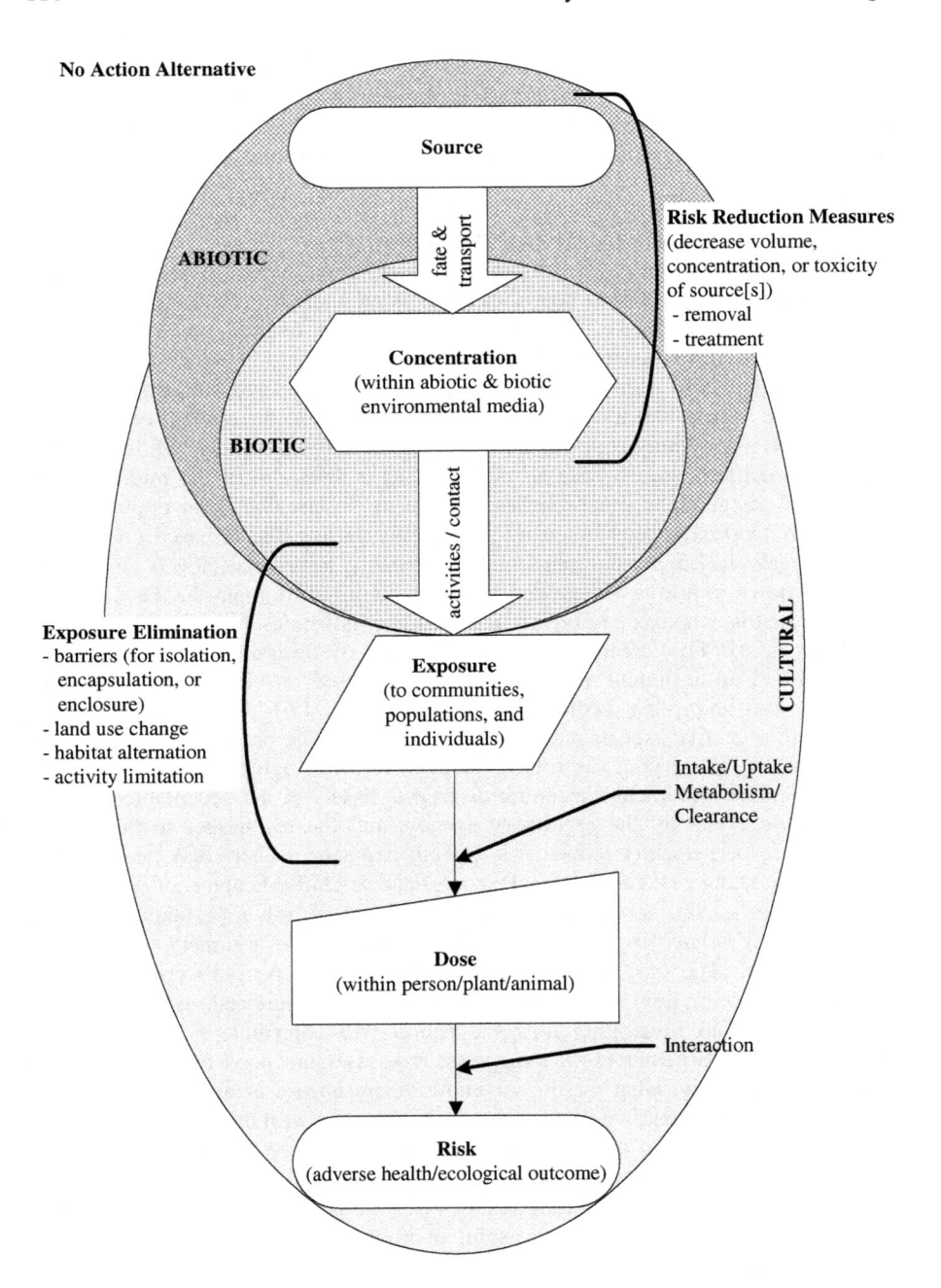

Figure 4.6 Controlling the Risk System.

ultimate remedy is optimized for actual site conditions (*e.g.*, optimized number, location, and pumping rate of extraction wells).

- Periodic review—of performance of the ongoing remedy and even the completed remedy (if it is temporary such as institutional control) should be evaluated on a regular basis to compare anticipated with actual results, to identify any potential deficiencies in the remedy's protectiveness, and to seek opportunities to improve long-term performance. This is especially important when the selected remedy relies on monitored natural attenuation.

- Improving remedy performance—through the assessment of performance monitoring data may be used to refine the remedy, such as modifying extraction rates or changing the pattern of extraction wells, for example. Such improvements are capable of shortening cleanup timeframes, thus reducing costs.

- Revisiting and modifying cleanup goals—may be necessary at some sites if performance data indicate that attainment of these objectives or levels is not technically practicable.

C. REMEDIAL OPTIONS
Many remedial options are available for implementation at a particular site, ranging from "no action" to wholesale removal, and there are many factors—*e.g.*, performance, cost, community acceptability—underlying the decision to select a particular remedial strategy (see Table D.16) from possible options such as:

- No Action
- Repair
- Operations and Maintenance Program
- Isolation / Institutional Controls
- Encapsulation
- Enclosure
- Removal and Disposal
- Removal and Treatment
- Treatment In-Place
- Natural Attenuation

D. VALUE CREATION
Alternatives to "end-of-the-pipe" control or remediation are worthy of exploration. Beginning with the end in mind, and seeking to maximize the value of the environmentally impaired property through redevelopment, the environmental manager can have a positive reason for action that balances the costs of remediation with the increased value of the improved property while mitigating health and ecological risks. In short, *redevelopment is remediation.*

We believe that reuse should drive remediation. With this mindset, the manager must begin making some difficult decisions before remediation. The decisions must be knowledge-based (appreciating both objective and subjective areas), informed by the work of different specialists, and include whether the site will be sold to a developer, subject to a long-term lease, and with what limitations ownership or use of the site will be transferred. The decisions, like many others, depend on site environmental conditions, local laws, community issues and needs, and liability considerations.

Additionally, the company must be able to balance opinions and stick to decisions. Redevelopment processes pit special interest groups against each other. Former friends of a company may become adversaries. The company must not overreact to negative inputs, but be committed to a strategy that allows it to maintain a directed approach to the endpoint—site reuse with risk mitigation that benefits the company, achieves regulatory buy-in, and which stakeholders see as positive.

Table 4.4 augments the conceptualization of Figure 3.7 with steps recognized by the planning profession as necessary considerations associated with traditional site development, but adapted to the special considerations posed to the environmental manager by redevelopment of impaired property. Table D.17 presents an additional set of objectives focused on repositioning impaired property in their real estate markets for potential sale and/or redevelopment as part of a strategy to control costs of environmental remediation and escalate socioeconomic health.

VI.   DEVELOPING RISK ARGUMENTS

A risk "argument" is not an argument in the pejorative sense. Rather, a risk argument is a convincing communications approach to achieve optimal, "mutual gain" solutions by integrating property value, environmental (*i.e.*, human health and ecological) risk or impairment, situational context, risk management options, and decision-making frameworks. The risk information and preferred risk management options are formulated within a communication program by which it is presented and, ultimately, built into an agreeable approach acceptable to all concerned parties.

Although the main portion of this section is contained in Appendix C, several points need stating here.

- It should already be obvious that we prefer an integrated approach that considers interactive decisioning within a systematic process.

- The process involves the environmental manager and appropriate members of their team, regulatory decision-makers, technical experts, and appropriate stakeholders.

- As discussed elsewhere, the company must develop an institutional approach and guidelines on performing risk communications and how the integration will come about.

- Early and regular dialog is necessary with those within the organization and with outside parties. To nurture an integrative concept be sure to have the risk assessor involved in such contacts to build within them an understanding of the perceptions and concerns of the stakeholders. This

Table 4.4
Market-Driven Planning Process

Asset Definition	Perhaps obvious, but the first question to ask about an impaired property one wishes to position for redevelopment is, "Are we looking at good real estate, and what can we do about it given its environmental constraints?" Taking stock of existing site conditions can be accomplished with a tool known as a Concise Environmental and Redevelopment Assessment (CERSA, Ackerman *et al.*, 1998). CERSAs are traditional environmental site assessments correlated to site market data, fully compliant with state and federal regulatory requirements, and focused to help determine how environmental data work with market opportunity data.
Market Understanding	Market and demographic studies of areas surrounding a site are crucial to the process leading to preliminary site reuse plans. Development options come from these plans and may include: commercial, residential, industrial, and recreational (green space) uses, or any combinations. Input from stakeholders identified during situation analysis should be considered in developing the plans.
Constraints Analysis	Traditionally, constraints may include local zoning ordinances preventing a specific type of development, permitting considerations, traffic and access, title issues, and other encumbrances. An additional constraint here is the environmental impairment (either real or perceived) itself. The challenge lies in coordinating opportunities and constraints for redevelopment in a manner that can minimize environmental remediation costs. The CERSA described above becomes a "Comprehensive" Environmental and Redevelopment Assessment in this phase. Using more definitive environmental investigation and market data, the comprehensive assessment supports site remediation costs estimating—within a certain margin of error—for specific end uses.
Alternatives Analysis	Based on the scenarios selected, the management team identifies cleanup endpoints; this requires close interaction with the risk assessment step (Figure 3.7). Focused assessments are necessary to consider each reuse scenario, not only from a baseline risk perspective but from a risk of remedy perspective as well. Remediation/restoration endpoints consistent with the planned reuse are then identified and cleanup levels calculated accordingly. The approach taken to defining what is clean must be risk-based, but also must address long-term liability, regulatory, and other stakeholder concerns. This is where the concept of redevelopment as remediation can be crucial, using development concepts to spur the innovation of cost-effective source removal, techniques to cut-off exposure routes, and land uses that eliminate exposure/risk while simultaneously building value into the real estate asset.
Integrate Remedial/ Corrective Action and Redevelopment Plans for Chosen Alternative	Finally, the team must integrate remedial/corrective action plans with redevelopment plans for the chosen alternative. The inherent objective is refinement of preliminary redevelopment plans in coordination with appropriate stakeholders. The focus is to evaluate each preliminary plan iteratively for its environmental and economic risk potential in relationship to potential remedial action and associated redevelopment alternatives for the site, thereby optimizing each plan to reflect the most current information about potential remedial response scenarios. Additionally, the team may need more comprehensive, detailed analyses for critical redevelopment issues (*e.g.*, roadway/infrastructure improvements). A detailed plan and strategy for permitting with respect to each preliminary redevelopment plan will be necessary. A key aspect of strategic planning concerns phasing in redevelopment activities to match environmental cleanup options. For example, getting site areas "cleared" for development may enable a site to begin generating revenue while other site areas undergo remediation activities.

dialog also should include appropriate public informational materials to encourage understanding of the risk system, ensure clarity of expectations, and develop trust.

- The manager must maintain the long-view in order to cope with the all-too-frequent missteps and start-up problems as to ensure investment and commitment of not only the organization, but stakeholders as well. This also will require an institutional learning process to revise and augment plans and guidelines as the process proceeds.

- It is important to see risk communication as dialog that aims to build win/win solutions by helping one see the deeper side of the other position and begins to inform those positions (Robinson, 2000). In so doing, one is able to establish overlaps, and obtain definitions from the stakeholders that the manager can agree with. This leads to a shared understanding, which can benefit agreement and ultimately implementation of the remedial/redevelopment plan.

As stated earlier in this book, we believe that process is substance and when resolving environmental impairments with an eye towards redevelopment/reuse the process becomes a message of environmental protection and stewardship that can reward the company with "green branding" and the community with positive gain for those "who <u>were</u> at risk!"

VII. REDEVELOPMENT AS REMEDIATION OR ENVIRONMENTAL AND ECONOMIC FUSION

In concluding Chapter 4, this section steps back from the focal point of environmental risk to see an extending site line that looks out to a process vision for a sustainable urban-industrial landscape. In other words, we see Risk-Based Analysis as a "credible conduit" through which dreams for a sustainable urban-industrial landscape can materialize. As discussed throughout this book, managing environmental risk to redevelop or restore impaired and/or stigmatized properties transcends an immediate order of "cookie-cutter" steps when it comes down to reckoning with critical factors for success. Environmental conditions (nature and extent), future-use potential, site ownership, area infrastructure, funding resources, cost recovery, regulations, risk and liability management, community relations, and expected return on investment are only a *few* of the complexities that need to be dealt with in an *integrated* fashion to achieve desired results. The dynamics of these and other site-specific issues warrant scenarios beyond traditional environmental and real estate business practices.

A. OVERHAULING CONVENTION

Conventional wisdom asserts that managing environmental risk employs two primary *financial gauges*:

- risk of environmental liability, and

- the risk associated from the investment in mitigating environmental liability and creating value through site reuse options.

Table 4.5

Types of Impaired Sites

Viable Sites	Marginal Sites	Upside-Down Sites
Tier 1. Institutional	Tier 1. Institutional	
Tier 2. Venture Capital	Tier 2. Venture Capital	
Tier 3. Local Capital	Tier 3. Local Capital	Limited or no Investor

Stemming from this basis—where the thinking is that location is the "end-all" and "be-all" of any type of real estate—three hierarchical brands of Brownfields or other environmentally impaired sites are typically identified (see Table 4.5).

1. Economically Viable Sites

Informally known as "low-hanging fruit" in the real estate market, these sites either have minor environmental challenges or situations wherein the economic rewards of redevelopment and reuse appreciably outweigh the cost of site cleanup as a line item in the redevelopment budget. Simply stated, these properties offer good opportunities from a pragmatic investment perspective and therefore attract institutional investment capital. Transactional deals are possible, with environmental issues being relatively minor inconveniences. If all sites fell into this category, corporate managers would have a relatively easy and straightforward job. However, the vast majority of environmentally impaired properties do not enjoy such economic veracity.

2. Marginal Sites

Again, from an established business perspective, the threshold of economic returns associated with such sites mandate that their redevelopment is not possible gainfully—at least in whole—without employing Risk-Based Analysis in an incentive-based setting. The difference between the costs of mitigating environmental liability and realizing the financial return from redevelopment makes these sites "too risky" for most conventional, private-sector investment applications. These sites often fall below the investment grades for institutional capital but above the means of small investors. Opportunistic venture capital organizations and (rarely) local capital resources, however, may be interested in select sites falling into this category. The manager therefore faces a considerably difficult task in formulating an environmental management approach that will be economically responsible.

3. Upside-Down Sites

Appraisals often perceive an inordinate amount of environmental liability and significantly limited value in terms of economic redevelopment prospects in upside-down sites. Both private and public investment sources may have a "hands-off" attitude or policy regarding these sites. Opportunistic investment capital is, in all but the most rare cases, simply nonexistent. For the manager, how can Risk-Based Analysis help identify economic parameters associated with environmental cleanup on these sites?

B. UPSIZING SITES: OUTSIDE-THE-BOX THINKING

The orientation of placing Brownfields and other environmentally impaired sites into one of these three "boxes" is that the perspective depletes an appreciation of them in terms of their value to other sites in the vicinity and the economic welfare of communities in that vicinity. That is, by identifying these sites only in segregated terms obscures their inter-relational values. However, this obscurity can be erased and replaced with a comprehensive vision for these sites, which most investors would deem marginal or upside-down. Environmental risk management options can be leveraged through innovative "site positioning strategies" to create value ("upsizing") for a corporation and community in unison with providing for a sustainable future. It may appear that the focus on this section is economic versus environmental, but "versus" it is not because these two components cannot be segregated. Economic boundaries constitute much of environmental management realities for a corporate manager and all other stakeholders. On some sites, for example, a corporation's entire worth could be absorbed in cleaning up a site— without success! In such cases, not only would an environmental legacy remain but also the economic viability of the corporation (to pay taxes, provide jobs, etc.) would vanish. These potential disasters are why Risk-Based Analysis is as much a socioeconomic—as it is an environmental—approach to a sustainable future.

C. NEW PARADIGM: VALUE CREATION

Most marginal or upside-down sites remain inert—mothballed in environmental, political, social, technical, legal, economic, and regulatory limbo—with such barriers precluding any site restoration or redevelopment prospects even before their conception. Thus, how can we breathe new life into these situations? How does Risk-Based Analysis work toward a sustainable urban-industrial landscape?

1. Identifies Barriers

Although each site has its own unique attributes, identifying barriers to revitalization potential is an initial, classic step that needs completion in virtually all cases. These barriers then can become components in an economic and environmental model to prepare for site revitalization. The following items are common barriers with marginal and upside-down sites.

a. Divergent Stakeholder Objectives

These issues primarily relate to what constitutes "reasonable" cleanup standards, a lack of incentive for moving forward, caretaker cost avoidance, and site reuse flexibility. What constitutes "reasonable" cleanup standards is often a fragmented component because the future end use for a site remains elusive without stakeholder buy-in. Such buy-in has an economic value requiring quantification. Without "some numbers," a lack of incentive for revitalizing the site grows—because the value is unclear, uncorroborated, or even underived.

b. Lack of Predictable Cleanup Funding

Without stakeholder consensus, it is next to impossible to identify standards for cleanup and revitalization and, as a result, any associated funding sources.

Alternatively, if one identifies such funding sources, stakeholder divergence may breed an uncertainty in ascertaining a credible amount of required funding. Divergent perceptions of the value, standards, reuse, and costs regarding site cleanup further impede the predictability and timeliness of this funding.

c. Lack of Uniformity in Environmental/Economic Perspectives

Many stakeholders of marginal or upside-down sites lack the appropriate expertise—fearing institutional controls, liability, and exposure to various kinds of risk, etc.—instead of addressing these issues as integrated aspects of the site redevelopment/reuse process. They are therefore unable to buy into the benefits of environmental risk management options. Without such understanding, many stakeholders insist on cleanup to overly conservative standards, resulting in gridlock.

d. Risk-Averse Influences

These influences insist on: cleaning up sites, again, to the most conservative standards, avoiding innovative cleanup technologies, and focusing on "single parcel" versus "multiple parcel" property issues. The central need is to identify cleanup endpoints, weigh risk among scientific, social, and ethical parameters, and partner with the community to help validate decisions in this regard. Multiple cleanup standards often are adjustable and responsive to management according to determined land uses over a time-horizon, where the standards or goals modify in a fashion responsive to land use changes. Once again, such "flexibility" is often lost because fragmented stakeholder groups generally lead to an absence of trust regarding the management of environmental risk.

e. Bureaucratic Processes versus Performance-Driven Processes

In many cases, the process is the problem. The bureaucratic process for all types of conveyances—including agency sign-offs—delays the progress of managing these sites. Even worse, protracted regulatory processes have no timetable for site reuse requirements.

2. Considers Economic Catalysts

Using case studies and the collective feedback of stakeholders associated with various projects regarding marginal or upside-down sites, Ackerman and Soler (2000) derived some guidance on how to accelerate property change and revitalization. The following "tenets" are applicable in most cases.

a. Align the Interests of All Stakeholders

Based on successful case studies, the challenge of enabling site transfer and revitalization in the most expeditious manner requires *integrated site planning and redevelopment.* This requires "reuse visioning" or orientation to create value beyond the surface through cleanup cost reductions and progress for community improvement. An integrated redevelopment approach capitalizes on cost savings and value creation through an early-in-the-process, real estate market analysis that dovetails concrete and measurable "highest and best" reuse options with feasible cleanup scenarios to promote *stakeholder-allied* endorsement for property reuse.

The alliances also provide a basis for accessing public capital, such as subsidies for infrastructure improvements, to further stimulate market interest in site acquisition and redevelopment.

b. Begin with the End

This book cites the saying "begin with the end in mind" several times. To work toward a sustainable urban-industrial landscape, "begin with the end" means making the real estate drive the deal. After all, we are looking at real estate, are we not, albeit real estate with inherent environmental issues? In conjunction with Risk-Based Analysis, feasibility assessments help the manager analyze the marketability of a marginal or upside-down site, determine its "highest and best use," and ascertain infrastructure improvements, which can be paramount to implementing that use. Existing or planned infrastructure improvements can result in extensive positive publicity that can virtually *pull* the market to the site as opposed to traditional real estate models that bring the site to the market. When we "begin with the end," we identify the "end use" "up front."

c. Pursue a Fully Integrated Process

This process begins with a full site characterization. A full site characterization includes: environmental, economic, market, demographic, community input, political climate, tax structure, and many other components. Property transfers are more expeditious if one evaluates the environmental issues in the context of the holistic set of real estate considerations required to effect change toward a sustainable urban-industrial landscape.

d. Reposition the Site to Decrease Cleanup Costs and Expedite Redevelopment

The results from a fully integrated site characterization process inspires direction in answering the following question for each environmentally impaired property: what do you have to do to the real estate to create higher value and cost savings for expediting its reuse and improving the community?

3. Benefits of Upsizing

There are five major benefits related to pursuing a marketing and value-creation paradigm for marginal and upside-down properties.

a. Real Estate Drives the Process

Real estate, not remediation, drives the property transfer and revitalization process. Stakeholders identify/secure end-users and develop master site plans before the cleanup process begins. Prospective purchasers, regulators, and other stakeholders collaborate on a fully integrated plan for remedy and reuse, including (as appropriate) institutional controls and end dates.

b. Proactive Remedies

A marketing and value-creation paradigm implements remedies that are compatible with the specific end-use(s) of properties. For example, some infrastructure improvements require remedial designs to accommodate new transportation-related needs. In such cases, remedial objectives may be to construct covers to support or enable specific uses (such as parking), allowing their construction on some of the most contaminated areas of a site and thereby cutting-off exposures.

c. Incentives

The marketing and value-creation approach generates incentives for stakeholders to support each other in securing regulatory approvals for cleanup/redevelopment plans. Many conflicting opinions are resolvable if key stakeholders in a given site restoration/redevelopment project could participate in the "upside" for deals that can generate a profitable scenario. For example, if the restoration of a contaminated property creates jobs—or green space to improve the quality of life in a community—stakeholders become recipients of the "upside" created.

d. Value is Created

Value creation for a site occurs by repositioning it to excite market and community interest. For example, a major activity in this component is the corporate manager seeking to add value to the real estate by securing subsidized funds from Brownfield initiatives or other public sources to help finance infrastructure needs and other beneficial public uses. Infrastructure improvements are then leveragable to market properties and secure contracts from master developers and end users. Value on a site also increases by obtaining entitlements and permitted zoning uses. This component includes working with municipalities on these issues to unlock value and attract end uses.

D. THE FUTURE IS IN THE PAST AND THE GAIN IS IN THE LOSS

Working toward a sustainable urban-industrial landscape—which in most instances means upsizing marginal and upside-down sites—requires creating value beyond the apparent reality. The issues managers address to achieve this goal are often very complex and require sophisticated strategies, combining environmental and economic models, to enable successful outcomes. The basis for these models lies in the use analysis for the property (which considers environmental risk as a constraint) and the property's impact on the surrounding vicinity. This use analysis is fluid in that it takes into account stakeholder interests in correlation with reasonable adaptive use options for subject properties. As a point of reference, turning a former factory site into a park has a total different set of economic and environmental influences if the site is in the middle of a residential neighborhood versus an industrial zone.

Table 4.6 offers a generic framework for these strategies and models. The hard work comes when applying this framework to a project-specific opportunity. Nevertheless, with genuine commitment from allied stakeholders, it really is possible to *reap the future from the past* (*i.e.*, sustainable economic, environmental, and social benefits created from a former industrial property with environmental impairment and stigma). More importantly perhaps for the corporate manager,

Table 4.6

Creating Value Beyond the Surface: A Generic Framework

Situational Challenges	Weighted Loss ($)	Value Creation Component	Weighted Gain ($)	Endpoint Value
Property Identified	Neutral	Exploratory Review of Site	Neutral	Reuse Process Begins
Stakeholder Discord	Negative	Stakeholder Input & Alignment	Positive	Site Use Consensus & Funding Leverage Begins
Lack of Risk Appraisal	Neutral	Problem Description & Definition	Neutral	Critical Information Gathered
Risk-Averse Impacts	Negative	Integrated Strategy: Risk & Rewards	Neutral	Seed Funding Sources Sought, Predevelopment Begins
Lack of Vision or Redevelopment Evaluation	Neutral	Site Repositioning through the Planning Process	Positive	Alliances & Defined End, then Attract Public Funding
Site Reuse Value Not Understood	Neutral	Defined Redevelopment Plan	Positive	Real Estate Value Created
Environmental Liability Burden	Negative	Transfer Risk of Loss	Positive	Liability Management With Cost Cap and Pollution Prevention Insurance
Lack of Financing	Neutral	Debt/Equity Structures for Capital Investment	Positive	Redevelopment Occurs with Public/Private Partnerships
Process Delays/Impacts	Negative	Performance Benchmarks Met to Accelerate Exit Strategy	Positive	Exit Strategy Completed & Site Revitalization Realized

business gains are realizable through loss (*i.e.*, repositioning the negative economic components of a property to derive positive value).

Using an environmental–economic model, one can ascribe a monetary value for each step of the site cleanup and reuse process, starting with site identification and review. Following an analysis to define and scope (formulate) the problem and create an integrated strategy, one formulates a baseline from which the financial profile and parameters for project success are determined. At each remaining step in the process, value rises above the baseline. The more value created, the more a sustainable urban-industrial landscape vision takes hold.

Appendix A

EVOLUTION OF THE RISK PARADIGM

Cris Williams

I. HISTORICAL CONTEXT

Risk assessment in relative terms is a fledgling discipline, having become formalized over the last 25 years or so. An early form of risk assessment came with the discovery of radioactivity at the turn of the century, prompting biologists and health physicists to examine the health impact of exposure to this new phenomenon (Ross, 1995). Engineers also contributed fundamental information for risk assessment as they assessed the safety of nuclear power plants, dams, chemical plants, and other large civil projects. According to Lehr (1990), writings about risk assessment date back about 3,000 years, although our present level of interest began only in 1960. It was then that concerns about radiation in the environment led to development of methods to quantify exposures from known sources. In the early 1970's, risk assessment was introduced as a discipline using scientific data to evaluate health risks quantitatively (Barnard, 1994). Quantitative risk assessment, advanced by the Food and Drug Administration in the 1970's to evaluate cancer risk under the diethylstilbestrol (DES) Proviso to the Delaney Clause, involved the use of statistical models as tools to extrapolate from relatively high-dose experimental data to the low doses that humans are normally exposed and to extrapolate across species. The Delaney Clause itself, which forbade the use of any cancer-causing chemical in processed foods, was passed by the United States Congress in 1958 in response to early studies by toxicologists who discovered that treating laboratory animals with high doses of chemical substances caused these animals to develop tumors.

The Comprehensive Environmental Response, Compensation, and Liability Act (CERCLA, or "Superfund") of 1980 established a national program for responding to releases of hazardous substances into the environment (USEPA, 1989a). The National Contingency Plan (NCP) is the regulation that implements CERCLA. Among other things, the NCP establishes the overall approach for determining appropriate remedial actions at Superfund sites. The overarching mandate of the Superfund program is to protect human health and the environment from potential threats posed by uncontrolled hazardous substance releases.

To meet this mandate, the USEPA developed a human health and ecological evaluation process as part of its remedial response program. Their process of gathering and assessing human health risk information is adapted from well-established chemical risk assessment principles and procedures first outlined by the National Academy of Sciences (see below). The first formal human health risk assessment guidance manual was known as the Superfund Public Health Evaluation Manual (USEPA, 1986c). After several years of Superfund program experience conducting risk assessments at hazardous waste sites, they updated the Public Health Evaluation Manual with the Human Health Evaluation Manual (USEPA, 1989a).

This guidance is currently used in the evaluation of hazardous waste and other sites in many states around the country (as well as internationally; see below).

Before USEPA's Superfund program was initiated, the Resource Conservation and Recovery Act (RCRA) was established (42 USC S/9 321 *et seq.* 1976). Unlike Superfund, which is designed to remedy the mistakes in hazardous waste management made in the past at sites that have been abandoned or where a sole responsible party cannot be identified, the RCRA Corrective Action Program was developed to encompass active, or soon to be active, facilities that are permitted or seek a permit to treat, store, or dispose of hazardous waste. As a condition for obtaining a RCRA operating permit, these active facilities are required to clean up contaminants that are released or have been released in the past.

Historically, before taking enforcement action against parties responsible for a hazardous waste site, USEPA was required to determine that an imminent and substantial endangerment to public health or the environment existed because of the site (USEPA, 1989a). Such a legal determination was known an "endangerment assessment." An endangerment assessment often was prepared as a study separate from the risk assessment. With the passage of SARA (the Superfund Amendments and Reauthorization Act) in 1986 and changes in Agency practice, the need to perform a detailed endangerment assessment as a separate effort from the risk assessment was eliminated. For administrative orders requiring a remedial design or remedial action, endangerment assessment determinations are now based on information developed in the site risk assessment. Elements included in the risk assessment conducted at a Superfund site during the RI/FS (Remedial Investigation/Feasibility Study) fully satisfy the informational requirements of the endangerment assessment. In 1985, USEPA produced a draft manual specifically written for endangerment assessment, the Endangerment Assessment Handbook. USEPA has determined that a guidance separate from the Risk Assessment Guidance for Superfund (Human Health Manual and Environmental Evaluation Manual) was not required for endangerment assessment, and, therefore, it was never finalized.

As a companion document to the Human Health Evaluation Manual, USEPA published its Environmental Evaluation Manual (USEPA, 1989b). USEPA's Framework for Ecological Risk Assessment (USEPA, 1992a) proposed principles and terminology for the ecological risk assessment process. From 1992 to 1994, the Agency focused on identifying a structure for a more formalized ecological risk assessment guidance document. Proposed ecological risk assessment guidelines were published for public comment in 1996 (61 FR 47552-47631, September 9, 1996). Current guidance, known as Guidelines for Ecological Risk Assessment (USEPA, 1998b), resulted from this formalization process (see below).

The following sections of this appendix provide more detail concerning the evolution of the risk assessment paradigm by tracing the development of human health and ecological risk assessment guidance in the United States and overseas. Both the National Academy of Sciences and the USEPA have been instrumental in this evolutionary process. A description of this evolution provides not only the historical context for risk assessment, but it also articulates that Risk-Based Analysis is soundly based on established science and policy principles to help the corporate manager implement these processes in an effective fashion to achieve corporate and stockholder goals, regulatory requirements, and stakeholder needs.

II.   THE NAS "RED BOOK"

In response to a directive from the Congress of the United States, the Food and Drug Administration and other government agencies contracted with the National Academy of Sciences to conduct a study of the institutional means for risk assessment (NRC, 1983). The result of this collaboration was the publication, in 1983, of the book *Risk Assessment in the Federal Government: Managing the Process*, a work commonly referred to as the NAS "Red Book." The book explored the relationship between science and public policy in the assessment of the risk of cancer and other adverse health effects associated with exposure of humans to toxic substances, in an attempt to delineate institutional mechanisms to foster a constructive partnership between science and government. These mechanisms were intended to ensure that government regulation rests on the best available scientific knowledge and preserves the integrity of scientific data and judgments in the unavoidable collision of the contending interests that accompany most important regulatory decisions.

The National Academy of Sciences formed the Committee on Institutional Means for the Assessment of Risks to Public Health to respond to the congressional directive. The Committee established three objectives:

- to assess the merits of separating the analytic functions of developing risk assessments from the regulatory functions of making policy decisions;

- to consider the feasibility of designing a single organization to do risk assessments for all regulatory agencies; and

- to consider the feasibility of developing uniform risk assessment guidelines for use by all regulatory agencies.

They were not interested in examining scientific issues or broad social policy questions. Rather, their more limited purpose was to examine whether altered institutional arrangements or procedures can improve regulatory performance.

One notable contribution arising from this effort to the evolution of the risk assessment paradigm was the distinction between risk assessment and risk management:

- risk assessment is the use of the factual base to define the health effects of exposing individuals or populations to hazardous materials and situations, and

- risk management is the process of weighing policy alternatives and selecting the most appropriate regulatory action by integrating the results of risk assessment and engineering data with social, economic, and political concerns to reach a decision.

This conceptual distinction is useful under certain circumstances, such as insulating scientific activity from political pressure and maintaining the analytic distinction between the magnitude of a risk and the cost of coping with it (NRC, 1996). The Red Book, however, recognized the limitations of a strict separation between risk assessment and risk management. This recognition pointed to the need to iterate between the two so that risk assessment could incorporate alternate assumptions for differing functions, such as initial screening or the evaluation of regulatory options (NRC, 1996). In this vein, the Red Book stated:

> *"Separation of the risk assessment function from an agency's regulatory activities is likely to inhibit the interaction between assessors and regulators that is necessary for the proper interpretation of risk estimates and the evaluation of risk management options. Separation can lead to disjunction between assessment and regulatory agendas and cause delays in regulatory proceedings."*

Remarkably, this recognition was largely ignored. The implied distinction between risk assessment and risk management inspired many risk assessors to believe that the two were separated by an inviolate "Chinese Wall." This practice "belief" led to paralysis and a frustrating process because risk assessors disavowed themselves from participation in risk management, effectively closing their ears to the needs of risk managers and others in decision-making. Fortunately, it also stimulated the evolution of the process.

The other notable contribution noted in the Red Book is the articulation of the four steps of risk assessment.

- **Hazard Identification.** This is the process of determining whether exposure to a chemical can cause an increased incidence of a particular adverse health effect (*e.g.*, cancer, organ toxicity), and whether the adverse health effect is likely to occur in humans.

- **Exposure Assessment.** This is the process of estimating, either qualitatively or quantitatively, the magnitude, frequency, duration, and route of exposure. In this case, exposure means contact by people with a chemical. Exposure is quantified as the amount of the agent available at the exchange boundaries (*e.g.*, skin, lungs, and gut) and available for absorption.

- **Toxicity Assessment.** This step in the process characterizes the relationship between the dose of a chemical received and the occurrence of adverse health effects in the exposed population. This process provides toxicity values for use in estimating risk.

- **Risk Characterization.** This step combines the toxicity and exposure assessments into quantitative expressions of risk. The exposure estimates are compared to chemical-specific toxicity values to determine the likelihood of adverse health effects in potentially exposed populations.

These steps continue to serve as the foundation for risk assessments performed today under a variety of regulatory regimes.

III. RISE OF THE RISK-BASED CORRECTIVE ACTION
CONCEPT

In the 1980s, to satisfy the need to start corrective action programs quickly, many petroleum underground storage tank (UST) implementing agencies decided to utilize regulatory cleanup standards developed for other purposes and apply them uniformly to UST release sites to establish cleanup requirements. With experience, however, it became increasingly apparent that applying such standards without consideration to the extent of actual or potential human and environmental exposure was an inefficient means of providing adequate protection against the risks associated with UST releases. In an attempt to streamline and standardize the process associated with petroleum site assessment and cleanup while still assuring protection of human health and the environment, UST implementing agencies began looking to a method known as "risk-based corrective action" or RBCA (commonly pronounced as "Rebecca"). USEPA's "formal" definition of RBCA is:

> "A streamlined approach in which exposure and risk assessment practices are integrated with traditional components of the corrective action process to ensure that appropriate and cost-effective remedies are selected, and that limited resources are properly allocated."
> (http://www.epa.gov/swerust1/rbdm/rbdmfaq6.htm)

RBCA is a tiered approach originally conceived by engineers at Shell Oil Company, who took their initial concepts to the ASTM to develop a national standard for assessing petroleum contaminated sites. In 1994, ASTM issued an emergency standard entitled *Guide for Risk-Based Corrective Action Applied at Petroleum Release Sites* [ES-38-94]. USEPA's March 1, 1996 OSWER 9610-17 Directive formalized the use of risk-based decision-making in federal and state UST corrective action programs. Under this directive, risk-based decision-making is a process that implementing agencies can use to make determinations about the extent and urgency of corrective action and about the scope and intensity of their oversight of corrective action by UST owners and operators. The real value of risk-based decision-making lies in its potential to help UST implementing agencies and UST owners and operators oversee/manage cleanups of UST releases based on relative risks to human health and the environment. In addition, risk-based decision-making can provide a coherent directional framework to help keep transaction costs under control. Thus, while risk-based decision-making can be as protective of human health and the environment as other approaches, it offers a more scientifically sound and administratively effective way to respond to the pressures for timely action at large numbers of sites and efficient use of both public and private resources.

Risk-based decision-making is a mechanism for identifying necessary and appropriate action throughout the corrective action process. Depending on known or anticipated risks to human health and the environment, appropriate action may include site closure, monitoring and data collection, active or passive remediation, contaminant, or institutional controls. In all cases, the objective is the same, *i.e.*, to ensure that adequate protection of human health and the environment is provided. The availability of options such as allowing contamination to remain in place or using institutional controls to prevent exposure will depend on applicable state and local laws and regulations.

The RBCA process recognizes the diversity inherent in petroleum and other chemical release sites and uses a tiered approach where corrective action activities are tailored to site-specific conditions and risks. Ecological risk assessment under RBCA is a qualitative evaluation of the actual or potential impacts to environmental (non-human) receptors.

The RBCA decision-making process integrates risk and exposure assessment practices, as suggested by the USEPA, with site assessment activities and remedial measure selection to ensure that the chosen action is protective of human health and the environment. The following general sequence of events is prescribed in RBCA (as outlined in the most recent ASTM standard for chemical release sites) once the process is triggered by the suspicion or confirmation of a release (ASTM, 1998 and 2000):

- performance of a site assessment;
- classification of the site by the urgency of initial response;
- implementation of an initial response action appropriate for the selected site classification;
- comparison of concentrations of chemical(s) of concern at the site with Tier 1 Risk Based Screening Levels (RBSLs) given in a look-up table;
- deciding whether further tier evaluation is warranted, if implementation of interim remedial action is warranted, or if RBSLs may be applied as remediation target levels;
- collection of additional site-specific information as necessary, if further tier evaluation is warranted;
- development of site-specific target levels (SSTLs) and point(s) of compliance (Tier 2 evaluation);
- comparison of the concentrations of chemical(s) of concern at the site with the Tier 2 evaluation SSTL at the determined point(s) of compliance or source area(s);
- deciding whether further tier evaluation is warranted, if implementation of interim remedial action is warranted, or if Tier 2 SSTLs may be applied as remediation target levels;
- collection of additional site-specific information as necessary, if further tier evaluation is warranted;
- development of SSTL and point(s) of compliance (Tier 3 evaluation);
- comparison of the concentrations of chemical(s) of concern at the site at the determined point(s) of compliance or source area(s) with the Tier 3 evaluation SSTL; and
- development of a remedial action plan to achieve the SSTL, as applicable.

RBCA principles have been adopted in many state programs to assess, in addition to petroleum and other chemical release sites, dry-cleaning sites, and

Brownfields (ASTM, 1998). Section XI below provides more detail concerning the use of RBCA is various state voluntary cleanup programs.

IV. THE NAS "BLUE BOOK"

The 1994 National Academy of Sciences (NAS) publication *Science and Judgment in Risk Assessment* (commonly referred to as the "Blue Book"; NRC, 1994) was developed in response to Section 112(o) of the 1990 Clean Air Act, which directs the USEPA to:

- review its methods to determine the carcinogenic risk associated with exposure to hazardous air pollutants from sources subject to Section 112;

- include in its review evaluations of the methods used to estimate the carcinogenic potency of hazardous air pollutants and for estimating human exposures to these pollutants; and

- evaluate, to the extent practicable, risk assessment methods for non-carcinogenic health effects for which safe thresholds might not exist.

Although risk assessment of hazardous air pollutants is the focus of the book, many of the issues discussed are applicable to all aspects of risk assessments. Further, many of these issues are still under debate.

One of the main themes touched upon in the book concerns USEPA's use of "default options." These options are used in the absence of convincing scientific knowledge on which several competing models and theories are correct. The options are not rules that bind the Agency; rather, they constitute guidelines from which the Agency may depart when evaluating the risk posed by a specific substance. For the most part, the defaults are conservative (*i.e.*, they represent a choice that, although scientifically plausible given existing uncertainty, is more likely to result in overestimating than underestimating risk). The book indicates that the Agency:

- often does not clearly articulate in its guidelines that a specific assumption is a default option;

- does not fully explain in its guidelines the basis for each default option;

- while allowing for departure from a default option in a specific case when it ascertains that there is a consensus among knowledgeable scientists that the available scientific evidence justifies the departure, no criteria exist guiding departures.

Another theme highlighted in the book concerns specific models and methods used by the USEPA to assess risks. The book recommends the Agency continue to explore and, where scientifically appropriate, incorporate pharmacokinetic modeling (*i.e.*, modeling the link between exposure and the biologically effective dose, or the dose that actually reaches the target tissue). It also recommends that they continue to use the linearized multistage model for assessing the carcinogenic potency of chemical substances, and that the Agency incorporate chemical mechanism of action and individual and population differences in susceptibility to assess non-carcinogenic effects.

The Blue Book also offers recommendations regarding USEPA's treatment of variability and uncertainty in risk assessments:

- Maintain a distinction between variability and uncertainty throughout the assessment process.
- Concerning variability specifically:
  - adopt a default assumption for differences in susceptibility among humans in estimating individual risks, and
  - assess risks to infants and children whenever it appears that their risks might be greater than those of adults.
- Concerning uncertainty:
  - develop guidelines for quantifying and communicating uncertainty as it occurs into each step of the risk assessment process, and
  - consider the uncertainties in each input value in an assessment, rather than determining only point estimates of risk.

These recommendations were adopted with the publication of USEPA's *Process for Conducting Probabilistic Risk Assessment* (USEPA, 1999b).

The final NAS Committee recommendation concerned the need for an iterative approach to risk assessment. This approach would start with relatively inexpensive screening techniques. For chemicals exceeding *de minimus* risk, generally defined as a risk of adverse health effects of one in a million or less, the assessor would apply more resource-intensive levels of data gathering, model construction, and model application. To guard against the underestimation of risk, screening techniques must err on the side of caution when there is uncertainty about model assumptions or parameter values.

The recommendations of the Blue Book to the USEPA are summarized below:

- generally retain a conservative, default-based approach;
- develop and use an iterative approach to risk assessment, and provide justification for its current defaults and establish a procedure permitting departures from the default options; and
- when reporting risk estimates to decision-makers and the public, present not only single, point estimates, but also the sources and magnitudes of uncertainty associated with those estimates.

Two of the three of these recommendations were adopted to some degree following publication of the Blue Book. For example, default assumptions still dominate the Agency's risk assessment guidance, as well as the guidance of the majority of state regulatory agencies. Justification for the agencies' use of default assumptions is generally lacking, as is a defined protocol for developing alternatives to the default assumptions. The recent publication of USEPA's (1999b) *Process for Conducting Probabilistic Risk Assessment* represents the federal government's attempt to depart from the single-value or "point estimate" approach to risk assessment and characterize uncertainty and variability in the risk assessment process. Probabilistic techniques now are to apply only to the exposure assumptions used in the risk assessment and not to the toxicity values (see Section X below).

V. UNDERSTANDING RISK: INFORMING DECISIONS IN A DEMOCRATIC SOCIETY

Previous endeavors of the National Academy of Sciences (NRC, 1983 and 1994) focused on the linking of risk science and inherent policies. Likewise, NAS's 1996 effort *Understanding Risk: Informing Decisions in a Democratic Society* (NRC, 1996) continued the exploration of how best to translate risk information into a more usable form by a risk manager and interested parties. A committee of 17 individuals from a variety of specialties including risk assessment, epidemiology, toxicology, ecology, public policy, economics, decision science, social science, public health, and law was convened to assess opportunities to improve the characterization of risk to inform decision-making and resolution of controversies over risk. To this end, the committee addressed the following technical issues:

- representing uncertainty;
- translating the outputs of conventional risk analysis into non-technical language; and
- elucidating the social, behavioral, economic, and ethical aspects of risk that are relevant to the content or process of risk characterization.

The committee believed it necessary to "reconceive" risk characterization in order to increase the likelihood of achieving sound and acceptable decisions. The committee envisioned a process in which the characterization of risk emerges from a combination of analysis and deliberation. They offered seven principles for implementing the process as discussed below.

A. DECISION-DRIVEN: INFORMING CHOICES, SOLVING PROBLEMS

The committee noted that risk analysis has been criticized as being of little help for decision-making, even when it adds to scientific knowledge. Effective risk characterization must accurately translate the best available scientific information about risk into a language non-specialists can understand and appreciate. Good risk characterization results from a process that gets the science right—*i.e.*, involves an adequate level of scientific inquiry and analysis—and also gets the right science—*i.e.*, directs the analysis to the most decision-relevant issues.

B. CONSIDER RELEVANT LOSSES, HARMS, OR CONSEQUENCES

Risk analyses are criticized for focusing on only a narrow set of outcomes, *e.g.*, certain human health hazards. The committee argued that the outcomes that should be considered relevant depend on the decision and should not be narrowly focused and decided *a priori*. Risk characterizations should address social, economic, ecological, and ethical outcomes, as well as consequences for human health and safety. These characterizations also should address outcomes for particular populations in addition to risks to whole populations, maximally exposed individuals, or other standard affected groups. Adequate characterization depends on incorporating the perspectives and knowledge of the spectrum of interested and affected parties from the earliest phases of the effort. Further, the breadth of analysis and the appropriate extent of involvement or representation required for satisfactory risk characterization remain situation-dependent.

C. APPLY AN ANALYTIC-DELIBERATIVE PROCESS

The success of risk characterization depends on systematic analysis that is appropriate to the problem, responds to the needs of the interested and affected parties, and treats uncertainties of importance to the decision process in a comprehensible way. Success also depends on deliberations that formulate the decision problem, guide analysis to improve the decision participants' understanding, seek the meaning of analytic findings and uncertainties, and improve the ability of the interested parties to participate effectively in the decision process. The analytic-deliberative process must have an appropriately diverse representation of the interested and affected parties, of decision-makers, and of specialists in risk analysis.

D. START WITH PROVISIONAL ASSESSMENT OF THE DECISION

Risk situations vary widely, and one process is not necessarily appropriate for all risk characterizations. Specifically, the level of effort that should go into risk characterization is highly situation-dependent. For many decisions, a simple, generic risk characterization procedure will do. An inflexible decision to use a narrow, routine, or non-participatory, analytic-deliberative process for risk characterization can undermine the decision-making process and capitalize on irrelevance.

E. NEED FOR PROBLEM FORMULATION

It is important for the organizations responsible for risk decisions to investigate whether there are or might be competing definitions of the risk problem. Risk characterization can be fairly straightforward if the interested and affected parties agree on which issues deserve analysis; if they do not agree, efforts should be made at the outset to engage those parties in deliberations about what should be analyzed.

F. A MUTUAL AND ITERATIVE PROCESS

Analysis and deliberation are complimentary and should be integrated throughout the process leading to risk characterization: deliberation frames analysis, analysis informs deliberation, and the process benefits from feedback between the two. The interplay between analysis and deliberation sometimes merits revisiting past decisions. Covering old ground can in many cases improve understanding.

G. DEVELOPING ORGANIZATIONAL CAPABILITY

To possess the full range of analytic-deliberative capabilities, organizations may need to make special efforts to train staff in such concepts as participatory deliberation, the integration of analysis and deliberation, and social and ethical risk. It also may involve acquiring analytical expertise in areas of ecological, economic, or ethical outcomes, disciplines not typically possessed by experts in human health risk.

As described in Section II above, *Understanding Risk: Informing Decisions in a Democratic Society* was written in response to the perceived distinction between risk assessment and risk management and the development of the practice adopted by many risk assessors that the *characterization* of risk—*i.e.*, the quantitation of risk—

and the *management* of risk are separate activities conducted by different individuals or groups. It also was written in an attempt to encourage more broad-based participation in the risk assessment process by involving all potential stakeholders and thus served as a precursor document to the Presidential/Congressional Commission on Risk Assessment and Risk Management's *Framework for Environmental Health Risk Management* (see below).

VI. FRAMEWORK FOR ENVIRONMENTAL HEALTH RISK MANAGEMENT

Like the NAS "Blue Book" previously described, the 1990 Clean Air Act (CAA) amendments were the basis for the formation of a Presidential/Congressional Commission on Risk Assessment and Risk Management that, in 1997, published the *Framework for Environmental Health Risk Management* (PCCRARM, 1997a and 1997b). This commission was formed to

> *"...make a full investigation of the policy implications and appropriate uses of risk assessment and risk management in regulatory programs under various Federal laws to prevent cancer and other chronic human health effects which may result from exposure to hazardous substances."*

A clear need to modify the traditional approaches used to assess and reduce risks emerged as a major theme from the commission's deliberations. According to the commission, traditional approaches rely on a chemical-by-chemical, medium-by-medium, risk-by-risk strategy. In so doing, they tend to focus attention on refining assumption-laden mathematical estimates of the small risks associated with exposures to individual chemicals, rather than on the overall goal of reducing risk and improving health status.

The commission sought to create a framework to guide investments of valuable public sector and private sector resources in researching, assessing, characterizing, and reducing risk. The framework was designed to set forth principles for making good risk management decisions and for actively engaging stakeholders in the process. It also defined a six-stage management process that could be scaled to the importance of a public health or environmental problem and that:

- enables risk managers to address multiple relevant contaminants, sources, and pathways of exposure, so that threats to public health and the environment can be evaluated more comprehensively than is possible when only single chemicals in single environmental media are addressed;

- engages stakeholders as active partners so that different technical perspectives, public values, perceptions, and ethics are considered; and

- allows for incorporation of important new information that may emerge at any stage of the risk management process.

The commission's framework was designed to help all types of risk managers— government officials, private sector businesses, individual members of the public—

make good risk management decisions. The six stages of the framework are shown in Figure 2.9 and discussed below.

A. DEFINING THE PROBLEM

The commission deems the problem/context stage as the most important step in the framework. It involves:

- identifying and characterizing an environmental health problem, or a potential problem, caused by chemicals or other hazardous agents or situations;

- putting the problem into its public health and ecological context;

- determining risk management goals;

- identifying risk managers with the authority or responsibility to take the necessary actions; and

- implementing a process for engaging stakeholders.

The commission considers all these steps to be important, but they may be conducted in different orders, depending on the particular situation. For example, when a state or federal regulatory agency is mandated to take the lead on a problem, the steps often will proceed in the order listed above, with the identity of the risk managers already clear, since the state or federal agency will have assumed that role from the start. On the other hand, if the group or individual discovering the problem is not in a position to be the risk manager or to characterize the problem, stakeholders might have to engage in a collaborative stakeholder process to identify risk managers with the requisite authority before the other steps can take place.

B. ANALYZING RISKS

To make an effective risk management decision, risk managers and other stakeholders need to know what potential harm a situation poses and how great is the likelihood that people or the environment will be harmed. The nature, extent, and focus of a risk assessment should be guided by risk management goals. The results of a risk assessment—along with information about public values, statutory requirements, court decisions, equity considerations, benefits, and costs—are used to decide whether and how to manage the risks.

Risk assessors should provide risk managers and other stakeholders with plausible conclusions about risk that can be made based on the available information, along with evaluations of the scientific weight-of-evidence supporting those conclusions and descriptions of major sources of uncertainty and alternative views.

The commission lists the important questions to be addressed when analyzing risk:

- Considering the hazard and the exposure, what is the nature and likelihood of the health risk?

- Which individuals or groups are at risk? Are some people more likely to be at risk than others?

- How severe are the anticipated adverse impacts or effects?

- Are the effects reversible?

- What scientific evidence supports the conclusions about risk? How strong is the evidence?

- What is uncertain about the nature or magnitude of the risk?

- What is the range of informed views about the nature and probability of the risk?

- How confident are the risk analysts about their predictions of risk?

- What other sources cause the same type of effects or risks?

- What contribution does the particular source make to the overall risk of this kind of effect in the affected community? To the overall health of the community?

- How is the risk distributed in relation to other risks to the community?

- Does the risk have impacts besides those on health or the environment, such as social or cultural consequences?

C. EXAMINING OPTIONS

This stage of the risk management process involves identifying potential risk management options and evaluating their effectiveness, feasibility, costs, benefits, unintended consequences, and cultural or social impacts. This process can begin whenever appropriate after defining the problem and considering the context. It does not have to wait until the risk analysis is completed, although a risk analysis often will provide important information for identifying and evaluating risk management options. In some cases, examining risk management options may help refine a risk analysis. Risk management goals may be redefined after risk managers and stakeholders gain some appreciation for what is feasible, what the costs and benefits are, and what contribution reducing exposures and risks can make toward improving human and ecological health. Stakeholders can play an important role in all facets of identifying and analyzing options. They can help risk managers:

- develop methods for identifying risk-reduction options;

- develop and analyze options; and

- evaluate the ability of each option to reduce or eliminate risk, along with its feasibility, costs, benefits, and legal, social, and cultural impacts.

D. MAKING A DECISION

During this stage of the framework, decision-makers review the information gathered during the analyses of risks and options to select the most appropriate solution. When the risk problem falls under the purview of a federal, state, or local regulatory authority, the regulatory agency makes the risk management decision. Consumers, manufacturers, and others responsible for wastes and pollution can also make socially important decisions to reduce or eliminate risks. A productive stakeholder involvement process can generate important guidance for decision-makers. Thus, decisions may reflect negotiation and compromise, so long as statutory requirements and intent are met. In some cases, win-win solutions are available that allow stakeholders with divergent views to achieve their primary goals. Involving stakeholders and incorporating their recommendations where possible reorients the decision-making process from one dominated by regulators to one that includes those who must live with the consequences of the decision. This not only fosters successful implementation, but also can promote greater trust in government institutions.

E. TAKING ACTION

Traditionally, implementation has been driven by regulatory agencies' requirements. Businesses and municipalities are generally the implementers. However, the chances of success are significantly improved when other stakeholders also play essential roles. Depending on the situation, action-takers may include:

- Health and other public agencies

- Community groups

- Citizens

- Businesses

- Industries

- Unions/workers

- Technical experts

These groups can help:

- develop and implement a plan for taking action;

- explain to affected communities what decision was made and why and what actions will be taken; and

- monitor progress.

F. EVALUATING RESULTS

At this stage of risk management, decision-makers and other stakeholders review what risk management actions have been implemented and how effective they have been. Evaluating effectiveness involves monitoring and measuring, as well as comparing the actual benefits and costs to estimates made in the decision-making stage. The effectiveness of the process leading to implementation also should be evaluated at this stage. Evaluation provides important information about:

- whether the actions were successful, whether they accomplished what was intended, and whether the predicted benefits and costs were accurate;

- whether any modifications are needed to the risk management plan to improve success;

- whether any critical information gaps hindered success;

- whether any new information has emerged that indicates a decision or a stage of the framework should be revisited;

- whether the framework process was effective and how stakeholder involvement contributed to the outcome; and

- what lessons can be learned to guide future risk management decisions or improve the decision-making process.

Tools for evaluation include environmental and health monitoring, disease surveillance, analyses of costs and benefits, discussions with stakeholders, and other research. Evaluation is critical to accountability and to ensure wise use of scarce resources. Monitoring health indices can be one method of evaluating whether risk management has been successful.

Despite the emphasis on broad-based stakeholder involvement in the risk assessment and risk management process as advocated by this and the NAS study *Understanding Risk*, in general practice, the goal of all-inclusive stakeholder involvement has yet to be reached.

VII. USEPA'S REMEDY SELECTION RULES OF THUMB

Risk assessment and risk management "rules of thumb" were published in 1997 as a part of an overall effort by USEPA to organize into a single document key principles, expectations, and best practices concerning the Superfund remedy selection process (*Rules of Thumb for Superfund Remedy Selection*, USEPA 1997c). In addition to risk assessment rules of thumb, the USEPA published in this document rules of thumb for remedial alternatives and groundwater response actions. The discussion that follows pertains solely to the risk assessment and risk management rules of thumb. As the title implies, this document is a "practical" guide to risk assessment and risk management as it is to be performed within the context of USEPA's Superfund program. Recommendations contained within draw upon and build from previously (*e.g.*, USEPA, 1989a) and concurrently (*Ecological Risk Assessment Guidance for Superfund: Process for Designing and Conducting Ecological Risk Assessments*; USEPA, 1997a; see below) published Superfund documents and provide a blueprint for how the USEPA and many state regulatory agencies currently conduct risk assessments.

A. RISK ASSESSMENT RULES OF THUMB
The following principles are specified in the risk assessment portion of USEPA's rules of thumb document:

1. Conceptual Site Model
The rules of thumb recommend that a well-defined conceptual site model (CSM) be developed in the earliest stages of the baseline risk assessment. The CSM is a three-dimensional "picture" of site conditions that illustrates contaminant sources, release mechanisms, exposure pathways, migration routes, and potential human and ecological receptors. The CSM documents current and potential future site conditions and is supported by maps, cross sections, and site diagrams that illustrate what is known about human and environmental exposure through contaminant release and migration to potential receptors. The CSM is initially developed during the scoping phase of the RI/FS and should be modified as additional information becomes available.

2. Exposure Pathways

It is recommended that all relevant exposure pathways related to the site (*e.g.*, direct ingestion, inhalation) be evaluated, for both current and reasonably anticipated future land uses as well as current and potential future groundwater and surface water uses.

3. Data Needs

The rules of thumb stipulate the collection of sufficient contaminant concentration data from each relevant medium to characterize adequately the nature and extent of contamination and to develop sound estimates of risk associated with each exposure pathway.

4. Site-Specific Risk Calculations

The following principles apply to site-specific risk calculations in the baseline risk assessment:

- Calculate the cumulative risks to an individual for chronic exposures, using reasonable maximum exposure (RME) assumptions by combining a statistically sound, arithmetic average, exposure-point concentration with reasonably conservative values for intake and duration.

- Use the most current toxicity values provided by the Integrated Risk Information System (IRIS) or the Health Effects Assessment Summary Tables (HEAST).

- Include estimates of risk for current and reasonably anticipated future land uses and potential future groundwater and surface water uses, without institutional controls. The baseline risk assessment is essentially an evaluation of the "no action" alternative (*i.e.*, an assessment of the risk associated with a site in the absence of any remedial action or control). While institutional controls do not actively clean up the contamination at a site, they can control exposure and, therefore, are considered limited action alternatives that may be evaluated during the remedy selection process.

- Include a discussion that identifies major sources of uncertainty or variability and their influence on the risk estimates. Probabilistic methods may aid in evaluating uncertainty at some sites.

5. Other Measures of Risk

The risk assessment rules of thumb state that other measures of risk (*e.g.*, central tendency) can be used to describe site risks more fully. However, RME risk generally should be the principal basis for evaluating potential risks at Superfund sites.

6. Exposed Populations

The risk analysis should clearly identify the population, or population subgroup (*e.g.*, highly exposed or susceptible individuals), for which risks are being evaluated.

7. Ecological Risk Assessment

The rules of thumb recommend the inclusion of ecological risk in the baseline risk assessment in order to support USEPA's mission to protect the environment. A screening ecological risk assessment generally should be conducted to identify those chemicals, media, and portions of the site requiring a more detailed study and analysis.

B. RISK MANAGEMENT RULES OF THUMB

The following rules of thumb are involved when making risk management decisions in the Superfund program.

1. Basis for Action

A response action is generally warranted if one or more of the following conditions are met:

- The cumulative excess carcinogenic risk to an individual exceeds 10^{-4} (using reasonable maximum exposure assumptions for either the current or reasonably anticipated future land use);

- The non-carcinogenic hazard index is greater than one (using reasonable maximum exposure assumptions for either the current or reasonably anticipated future land use);

- Site contaminants cause adverse environmental impacts; or

- Chemical-specific standards or other measures that define acceptable risk levels are exceeded and exposure to contaminants above these acceptable levels is predicted for the RME (*e.g.*, drinking water standards that are exceeded in groundwater when that groundwater is a current or potential source of drinking water; or water quality standards that are exceeded in surface or groundwater that support the designated uses of these waters.

2. Preliminary Remediation Goals—Carcinogens

In the absence of ARARs for chemicals that pose carcinogenic risks, PRGs generally should be established at concentrations that achieve 10^{-6} excess cancer risk, modifying as appropriate based on exposure, uncertainty, and technical feasibility factors.

3. Preliminary Remediation Goals—Non-carcinogens

In the absence of ARARs for chemicals that pose non-carcinogenic risks, PRGs generally should be established at concentrations that achieve a hazard quotient of one. Cumulative non-cancer risks are determined by adding hazard quotients for chemicals with the same toxic endpoint or mechanism of action. In establishing PRGs for chemicals that affect the same target organ/system, PRGs for individual chemicals should be divided by the number of chemicals present in this group.

4. Chemical-Specific ARARs

When a single ARAR for a specific chemical (or in some cases a group of chemicals) defines an acceptable level of exposure, compliance with the ARAR generally will be considered protective even if it is outside the risk range (unless there are extenuating circumstances, such as exposure to multiple contaminants or pathways).

5. Background Concentrations

USEPA does not generally clean up below natural background levels. However, where anthropogenic (*i.e.*, man-made) background levels exceed acceptable risk-based levels, and USEPA has determined that a response action is appropriate, USEPA's goal is to develop a comprehensive response to address area-wide contamination. This "relative background perspective" will help avoid response actions that create "clean islands" amid widespread contamination.

6. Selecting Remedial Action

In the absence of ARARs, remedies should reduce the risks from carcinogenic contaminants so that the excess cumulative individual lifetime cancer risk or site-related exposures fall between 1×10^{-4} and 1×10^{-6}. The Agency has expressed a preference for cleanups achieving the more protective end of the risk range (*i.e.*, 1×10^{-6}). The upper boundary of the risk range is not a discrete line at 1×10^{-6}, although USEPA generally uses 1×10^{-4} in making risk management decisions. A specific risk estimate around 10^{-4} may be considered acceptable if justified based on site-specific conditions. For non-carcinogens, remedies generally should reduce contaminant concentrations so that exposed populations or sensitive sub-populations will not experience adverse effects during all or part of a lifetime, incorporating an adequate margin of safety (*i.e.*, a hazard index at or below one).

7. Timing

A "phased approach" to site investigation and cleanup generally will accelerate risk reduction and provide additional technical site information on which to base long-term risk management decisions. Phased cleanup approaches should be employed wherever practicable.

VIII. ECOLOGICAL RISK ASSESSMENT GUIDANCE

Ecological Risk Assessment Guidance for Superfund: Process for Designing and Conducting Ecological Risk Assessments (USEPA, 1997e) provides guidance on the process of designing and conducting ecological risk assessments for the Superfund Program. It is intended to promote consistency and a science-based approach within the Program, based on the *Proposed Guidelines for Ecological Risk Assessment* (USEPA, 1996a) and the *Framework for Ecological Risk Assessment* (USEPA, 1992a) developed by the Risk Assessment Forum of the USEPA (see Exhibit I-1 of http://www.epa.gov/oerrpage/superfund/programs/risk/ecorisk/intro.pdf).

Ecological risk assessment is an integral part of the RI/FS process, which is designed to support risk management decision-making for Superfund sites. The RI component of the process characterizes the nature and extent of contamination at a hazardous waste site and estimates risks to human health and the environment posed by contaminants at the site. The FS component of the process develops and evaluates remedial options. Thus, ecological risk assessment is fundamental to the RI and ecological considerations are part of the FS process.

This ecological guidance is intended to facilitate defensible site-specific assessments. Ecological risk assessment is an evolving technique, and this guidance represents a dynamic process framework that may change as assessment approaches improve. Thus, it does not dictate the scale or complexity of an assessment nor direct the user to specific protocols or investigation methods. Rather, professional judgment is emphasized in designing and determining the data needs for an assessment. However, when the process outlined in this document is followed, a technically defensible and appropriately scaled site-specific ecological risk assessment should result. This document supersedes the USEPA's (1989c) *Risk Assessment Guidance for Superfund, Volume II: Environmental Evaluation Manual.* However, the *Environmental Evaluation Manual* contains information on the statutory and regulatory basis of ecological assessment, basic ecological concepts, and other background information that is not repeated in the guidance.

The *Framework* is similar to the National Research Council's (NRC) paradigm for human health risk assessments (NRC, 1983; see above and Figure A.1). The 1983 NRC paradigm consists of four fundamental phases: hazard identification, dose-response assessment, exposure assessment, and risk characterization. The *Framework* differs from this paradigm:

- Problem formulation is incorporated into the beginning of the process to determine the focus and scope of the assessment.

- Hazard identification and dose-response assessment are combined in an ecological effects assessment phase; and

- The phrase "dose-response" is replaced by "stressor-response" to emphasize the possibility that physical changes (which are not measured in "doses") as well as chemical contamination can stress ecosystems.

Moreover, the *Framework* emphasizes the parallel nature of the ecological effects and exposure assessments by joining the two assessments in an analysis phase between problem formulation and risk characterization.

The guidance consists of eight steps and several scientific/management decision points (SMDPs; see Table A.1). A decision point requires a meeting between the risk manager and the assessment team to evaluate and approve or redirect the work up to that point. This group decides whether the assessment is on course. The SMDPs include a discussion of the uncertainty associated with the risk assessment that might be reduced, if necessary, with increased effort. These decision points are significant communication points that should be augmented with the consensus of all involved parties.

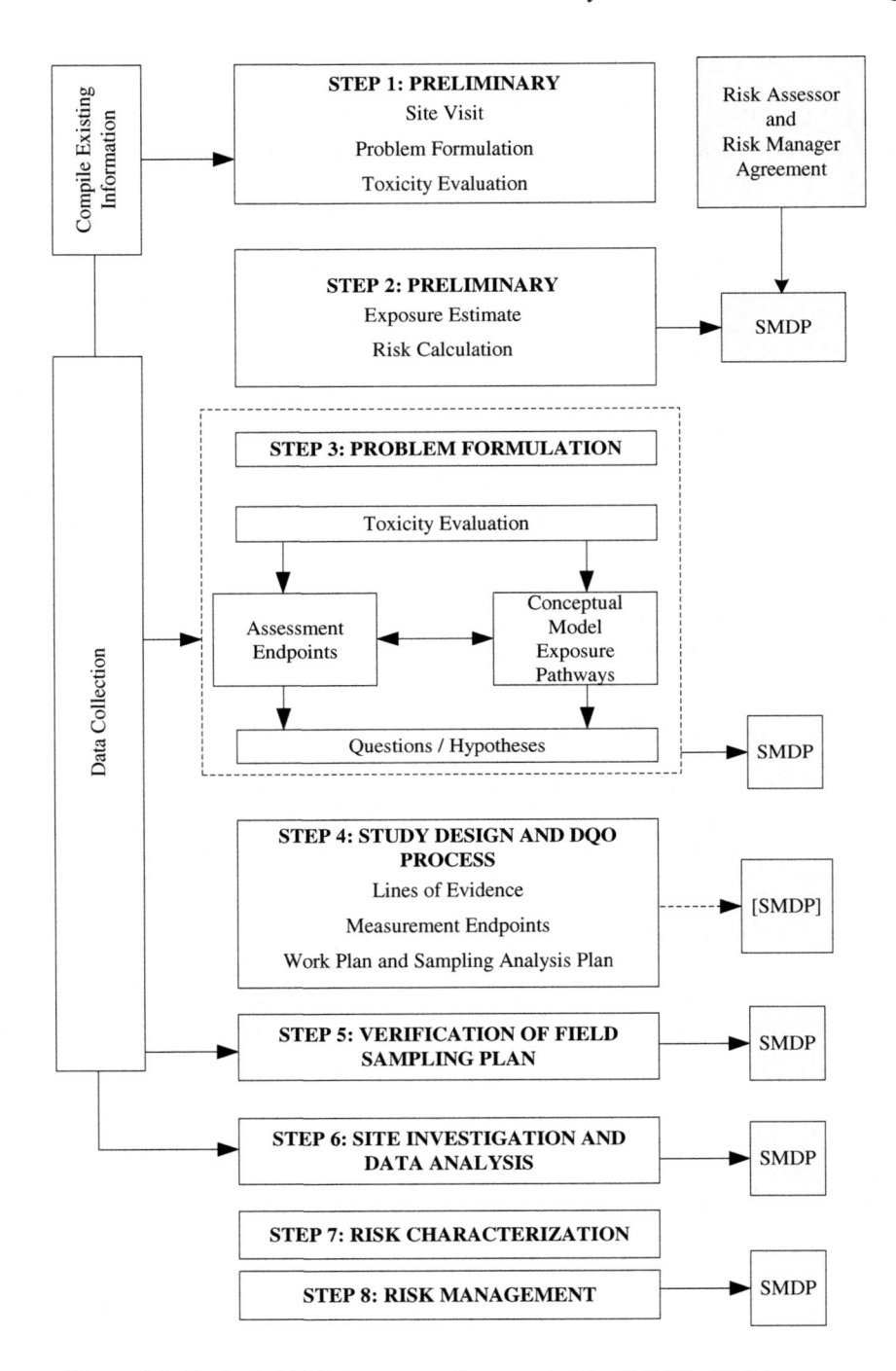

Figure A.1 Ecological Risk Assessment Framework (after USEPA, 1992a).

Table A.1

Ecological Risk Assessment Process Steps
Tied to Corresponding Superfund Process Decisions

Steps and Scientific/Management Decision Points (or SMDPs)

1.	Screening-Level Problem Formulation and Ecological Effects Evaluation	
2.	Screening-Level Preliminary Exposure Estimate and Risk Calculation	SMDP (a)
3.	Baseline Risk Assessment Problem Formulation	SMDP (b)
4.	Study Design and Data Quality Objectives	SMDP (c)
5.	Field Verification of Sampling Design	SMDP (d)
6.	Site Investigation and Analysis of Exposure and Effects	[SMDP] ()
7.	Risk Characterization	
8.	Risk Management	SMDP (e)

Corresponding Superfund Process Decision Points

(a)   Decision about whether a full ecological risk assessment is necessary.

(b)   Agreement among the risk assessors, risk manager, and other involved parties on the conceptual model, including assessment endpoints, exposure pathways, and questions or risk hypotheses.

(c)   Agreement among the risk assessors and risk manager on the measurement endpoints, study design, and data interpretation and analysis.

(d)   Signing approval of the work plan, and sampling and analysis plan for the ecological risk assessment.

(e)   Signing the Record of Decision.

[SMDP]   only if changes to the sampling and analysis plan is necessary.

IX.   COMMUNITY INVOLVEMENT

Involving the community in decisions concerning hazardous waste and other sites has recently become a priority at USEPA. Community involvement "in action" is typified by the USEPA's *Federal Facilities Stakeholder Involvement: Blueprint for Action* (USEPA, 1999a). This Blueprint was stimulated by USEPA's experience that demonstrated that cleanup at federal facilities improves when local stakeholders share information and participate in environmental decision-making.

Stakeholders are defined in the Blueprint as:

- local communities and governments;
- tribal communities and governments;
- civic and labor organizations;
- environmental justice groups;
- local redevelopment boards;
- educational institutions;
- state agencies;
- federal agencies; and

- individual citizens.

The Blueprint outlines a comprehensive program to ensure stakeholder involvement by putting citizens first. The Blueprint charts the direction for all stakeholders in four essential areas, as discussed below.

A. DIALOGUE

Dialogue is exemplified by the Federal Facilities Environmental Restoration Dialogue Committee, a forum established by USEPA to create a standard for public participation and develop a model for all serious environmental dialogues. The committee members released a report in 1996 that outlined a series of principles and recommendations reflecting the consensus of those involved with and affected by federal facilities cleanup decisions.

B. PARTNERSHIPS

USEPA recognizes that successful cleanup programs depend on strong partnerships. USEPA's Federal Facilities Restoration and Reuse Office (FFRRO) joins with the Department of Defense (DOD), the Department of Energy, and other stakeholders to facilitate faster, more effective, and less costly cleanups. For example, USEPA's FFRRO partners with DOD to implement the Fast-Track Cleanup Program. The intent of this program is to accelerate cleanups and speed the economic recovery of communities affected by military base closures.

C. ENVIRONMENTAL JUSTICE

Because environmental benefits and burdens may not be distributed equally throughout the population, USEPA has made a fundamental change in the way it implements environmental decisions. FFRRO integrates Environmental Justice perspectives into its work.

D. STAKEHOLDER INVOLVEMENT

USEPA recognizes that each citizen has a stake in the future of federal facilities. Consequently, FFRRO involves citizens groups, tribal parties, and state and local agencies in the cleanup process. Stakeholder involvement is implemented through the provision of resources, information, and training.

USEPA has recently developed its Community Based Environmental Protection (CBEP) program (http://www.epa.gov/ecocommunity/about.htm). Traditionally, environmental protection programs have focused on a particular medium or problem (*i.e.*, a "Command and Control" approach to environmental protection). These "Command and Control" programs have been effective at reducing point source pollution and improving environmental quality over the past two-and-one-half decades. However, some environmental problems, such as non-point source pollution, which may involve several media types and diffuse sources, are less amenable to "Command and Control" programs. Instead, a solution that seeks to address the various causes of the problems and understand the interrelationships between human behavior and pollution in a specific area may be more appropriate. CBEP supplements and complements the traditional environmental protection approach by focusing on the health of an ecosystem and the behavior of humans that live in the ecosystem's boundaries, instead of concentrating on a medium or particular problem. Therefore, CBEP is place-based, and not media or issue-based.

The CBEP approach has several other qualities that complement and supplement traditional environmental protection. Under a place-based protection scheme, the number and diversity of stakeholders tends to increase. For example, where an air pollution program may bring together a few industry representatives and special interest groups interested in air quality, a place-based program affects all individuals, groups, and industries concerned with the health and sustainability of a certain geographic area. Collaboration between diverse public and private stakeholders within a specific geographic area facilitates:

- comprehensive identification of local environmental concerns;
- the setting of priorities and goals that reflect overall community concerns; and
- the forging of comprehensive, long-term solutions.

The CBEP approach also connects and broadens the issues dealt with by environmental protection programs. Often a particular environmental problem, such as non-point source pollution, is affected by and related to several other environmental and resource issues in a geographic area. In order to solve one environmental problem, the related and connected environmental concerns also must be addressed.

Additionally, the CBEP approach recognizes the place of humans in ecosystems. Therefore, human economic and social needs must be developed in concert with environmental solutions to promote a sustainable future. A place-based focus allows stakeholders to identify the interrelated problems and forge a comprehensive, long-term plan that addresses the needs of the environment and its citizens. Therefore, the CBEP approach to environmental protection is holistic, not linear and isolated.

Finally, the CBEP approach can improve environmental program management. A large, diverse group of stakeholders can provide a wide array of expertise and knowledge when properly informed of an area's interrelated problems. This encourages the development of effective and appropriate problem-solving tools. Widespread stakeholder collaboration also improves environmental protection management by providing a means and forum for adaptive problem solving. If a problem-solving method is not working, the relationships established under collaborative work should facilitate discussion and implementation of alternative approaches. Therefore, the CBEP approach, by tapping into a high level of expertise and collaborative relationships, is an effective management tool.

X. PROBABILISTIC RISK ASSESSMENT

USEPA Region III published guidance in the form of a series of recommendations on the use of Monte Carlo simulation techniques in risk assessment in 1994 (http://www.USEPA.gov/reg3hwmd/risk/guide1.htm). The recommendations suggested how the technique is to be used in conjunction with conventional "point-estimate" risk assessments:

- Only human considered, environmental receptors are excluded.

- Work Plans must be submitted for USEPA review prior to the work to ensure the work will be acceptable to USEPA, and should describe the software to be used, the exposure routes and models, and input probability distributions and their sources.

- Only the exposure variables of the risk equation are considered probabilistically; noncancer reference doses and carcinogenic slope factors are considered as single numbers ("point estimates").

- Only significant exposure scenarios and chemicals are to be included, *i.e.,* exposure routes for which the risk exceeds either 1×10^{-06} cancer risk or a non-carcinogenic Hazard Index of 1.0, and chemicals that contribute 1% or more of the total risk or hazard index.

Likewise, in 1995, USEPA Region VIII developed a guidance document entitled *Use of Monte Carlo Simulation in Performing Risk Assessments* (USEPA, 1995c), following Region III; this guidance required a "point-estimate" approach to accompany any probabilistic analysis.

Such guidance was largely supplanted by the most recent installment in USEPA's *Risk Assessment Guidance for Superfund* (RAGS), which is *Volume III: Process for Conducting Probabilistic Risk Assessment* (USEPA, 1999b). This document describes what a probabilistic risk assessment (PRA) is and compares and contrasts it to the more familiar point estimate methods described in USEPA's RAGS Volume I (USEPA, 1989b). A risk assessment performed using probabilistic methods is very similar in concept and approach to the traditional point estimate method. The main difference is the methods used to incorporate variability and uncertainty into the risk estimate.

A variety of modeling techniques can be used to characterize variability and uncertainty in risk. This guidance focuses on Monte Carlo analysis (MCA), which is one of the most common probabilistic methods. At some sites, probabilistic analysis may provide a more complete and transparent characterization of the risks and uncertainties in risk estimates than would otherwise be possible with a point estimate approach. Developing or reviewing a PRA may involve additional time and resources, and a PRA is not necessary or desirable for every site. The USEPA uses a tiered approach to determine if PRA is appropriate at a specific site.

PRA is a general term for risk assessments that use probability models to represent the likelihood of different risk levels in a population (*i.e.,* variability) or to characterize uncertainty in risk estimates. In human health risk assessments, probability distributions for risk reflect variability or uncertainty in exposure. In ecological risk assessments, risk distributions may reflect variability or uncertainty in exposure or toxicity. A PRA that evaluates variability can be used to address the question, "What is the likelihood (*i.e.,* probability) that risks to an exposed individual will exceed a regulatory level of concern?" For example, based on the best available information regarding exposure and toxicity, a risk assessor might conclude, "It is estimated that there is a 10% probability that an individual exposed under these circumstances has a risk exceeding 1×10^{-6}." If a probabilistic approach also quantifies uncertainty, the output from a PRA can provide a quantitative measure of the confidence in the risk estimate. For example, a risk assessor might conclude, "While the best estimate is that there is a 10% chance that risk exceeds 1 x 10^{-6}, I am reasonably certain (95% sure) that the chance is no greater than 20%."

In the point estimate approach, a single numerical value (*i.e.*, point estimate) is chosen for each variable in the mathematical equation used to quantitate exposure and risk. For example, point estimates may include a drinking water ingestion rate of 2 liters per day and a body weight of 70 kilograms for an adult. Based on the choices that are made for each individual variable, a single estimate of risk is calculated. In the probabilistic approach, inputs to the risk equation are described as random variables (*e.g.*, variables can assume different values for different people) that can be defined mathematically by a probability distribution. For continuous random variables, such as body weight), the distribution may be described by a probability density function (PDF), whereas for discrete random variables (*e.g.*, number of fish meals per month), the distribution may be described by a probability mass function (PMF). The essential feature of PDFs and PMFs is that they describe the range of values that a variable may assume, and indicate the relative likelihood (*i.e.*, probability) of each value. For example, drinking water ingestion might be characterized by a normal distribution with a mean of 2 liters per day and a standard deviation of 1 liter per day. After determining appropriate PDF types and parameter values for selected variables, the set of PDFs are combined with the toxicity value in the exposure and risk equations given above to estimate a *distribution* of risks.

At this time, for human health risk assessments, toxicity values will generally be characterized by point estimates because of limitations in the data and techniques for characterizing distributions for toxicity in humans. Only if adequate supporting data are available to characterize variability or uncertainty in toxicity values will the Agency consider the use of distributions for toxicity. The Agency will determine the adequacy of supporting data on a case-by-case basis, pending consultation with USEPA Headquarters (Office of Emergency and Remedial Response). For ecological risk assessment, toxicity values may be characterized by probability distributions.

Both point estimate and probabilistic approaches can provide useful information for risk characterization. However, there are advantages and disadvantages associated with both methods that should be weighed before choosing to conduct a PRA. A point estimate approach should generally be performed before considering a PRA. If there is a clear value added from performing a PRA, then the use of PRA as a risk assessment tool may be considered.

By relying on the full scope of available information, PRA can often provide a more complete characterization of risk, as well as a quantitative description of the uncertainties in risk estimates. However, PRA generally involves additional effort throughout the risk assessment process and may not be needed for risk management at every site. Not all PRAs will involve the same level of effort to provide useful information for risk management decisions; a tiered approach, as described in the following paragraphs, is recommended to determine the appropriate level of analysis. In addition, the potential for misinterpretation of methods and conclusions is generally increased in PRA due to the greater complexity of the analysis.

A tiered, or stepwise, approach to PRA is advocated by USEPA. Tiered approaches to undertaking PRA have been discussed in the past and are commonly used for ecological risk assessment (USEPA, 1997b). The level of analysis and sophistication of methods used to quantify variability and uncertainty in exposure and toxicity can vary in complexity, depending on site-specific requirements. A tiered approach begins with a relatively simple analysis and progresses stepwise to analyses that are more complex. The level of complexity should match the site-specific risk assessment and risk management goals.

The initial steps of every PRA will generally involve a point estimate risk assessment. If the point estimate(s) of risk is/are greater than the level of regulatory concern, a risk assessor (together with other stakeholders) may consider whether or not the existing information will support a remedial decision, or whether additional risk assessment activities are warranted. At several points in the tiered approach, a question is posed, "Are the risks a concern?" To address this question, a risk assessor (and stakeholders) will generally consider the likelihood that the risk estimate exceeds a target risk level. It may also be important to consider the confidence in the risk estimates; that is, a risk estimate may be above or below a risk target, but judgment will be needed to determine the level of confidence that this risk estimate is sufficiently protective. The risk may be a concern if additional information on variability or uncertainty could lead to a different decision regarding remedial action. If additional probabilistic analysis is unlikely to make a difference in the risk management decision, then a decision generally should be made not to continue further with the tiered process for PRA.

Additional risk assessment activities should generally include an initial sensitivity analysis. This recommendation represents Tier 2 of the PRA and may be either a qualitative or a quantitative analysis depending on the complexity of the risk assessment at this point. For example, incidental ingestion of soil by children is often an influential factor in determining risk from soil, a fact recognized by risk assessors. This recognition is a *de facto* informal sensitivity analysis. A quantitative sensitivity analysis can also be performed to identify those exposure variables with the greatest influence on risk estimate.

If uncertainty in important variables can be quantified, then modeling approaches that separately characterize variability and uncertainty should be considered. This turning point represents Tier 3 of the PRA. Quantitative assessment of uncertainty and variability is also known as "Two-Dimensional" (2-D) PRA. Commercially available software (*e.g.*, Crystal Ball) supports this type of analysis.

PRA as it applies to contaminated site risk assessment today is in its infancy. At the time of this writing, USEPA's PRA guidance is in draft form. Once the guidance becomes final, PRA will slowly become a more common risk assessment technique. Remedial Project Managers in USEPA Region IV have stated that PRA has not played a role to date in cleanup decisions at sites in the Southeast United States (Pope, 2000). Over the next several years, it is predicted that PRA will become a standard technique for evaluating exposures that fail deterministic (*i.e.*, point estimate) screening assessments (Price and Keenan, 1997). It also is believed (Price and Keenan, 1997) that PRA will have a major impact on cost-benefit analyses that are part of evolving risk-based legislation in the United States Congress (http://www.cnie.org/nle/rsk-1.html).

XI.   PERSPECTIVES FROM STATE INITIATIVES AND VOLUNTARY CLEANUP PROGRAMS

State initiatives and voluntary programs concerning cleanup of contaminated sites have come into existence in recent years. Some of these programs, most notably the RCRA Underground Storage Tank Cleanup Program and special state programs for cleanup of dry cleaning facilities, rely heavily on voluntary compliance and may offer financial assistance for cleanup. For example, funds may be available to reimburse property owners for cleanup costs more than a certain amount. In addition, some property owners may elect to clean up their properties independently, without regulatory oversight.

Voluntary cleanup programs are state-sponsored programs that encourage private parties to clean up contaminated properties without enforcement by the state. They typically include requirements for eligibility, cleanup standards, and provisions for overseeing the cleanups. Most of these programs rely on volunteers to propose a cleanup plan, with the state typically reviewing and approving the plan. Forty-eight states allow volunteers to clean up contaminated property with some type of state review or approval (ELI, 1999). Only North Dakota, Wyoming, the District of Columbia, and Puerto Rico have no system for voluntary cleanups.

State programs vary considerably in how they approach voluntary cleanups. The cleanup standards or guidelines a state uses for deciding what amount of cleanup is required at sites are a large factor in determining the cost and length of cleanups. Most states have now established "risk-based" concentrations that stipulate how much contamination can be present on the site following cleanup. Although there are many different ways to establish these concentrations, two methods are far more commonly used than any others are. In the first, the state announces actual maximum concentrations for specific contaminants that can be allowed in soil or groundwater after a cleanup. These numerical values—often called generic standards or statewide health standards or default standards—are applicable to any site. In the second approach, the state allows the volunteer proposing the cleanup to develop contamination concentrations expressly for that site based on specific information about the contaminants present, the site's geologic characteristics, the potential use of the site, and other factors. The concentrations derived from these site-specific factors are an alternative way to establish maximum allowable concentrations of contaminants that meet the risk levels set by the state. It is important to note that many states allow parties to choose either method, perhaps another, or even to use a combination of methods.

In recent years, most states have decided to consider the future use of a site in setting cleanup standards. If a site will be used for an industrial or commercial facility—where children will not be exposed to contaminated soils, or groundwater will not be used for drinking—the cleanup standard may be set at levels that allow contaminated groundwater or soils to be left in place. This endpoint is considered acceptable because the planned land use of the site will reduce the risks that people will be exposed to while providing other social and/or economic enhancements. In such cases, so-called institutional controls may be used to assure that the use remains the same in the future and to protect public health and the environment if a future owner proposes to change the use of the site. Institutional controls are legal and administrative mechanisms that provide an additional method of reducing the likelihood of exposure by changing people's behavior so they avoid being exposed. Institutional controls include:

- warning signs;

- legal notices;

- land use controls and zoning;

- restrictions on how property may be used, often included in the deed to the property;

- restrictions on the use of groundwater for drinking; and

- warnings to people not to eat fish caught in particular lakes and streams, and education programs warning of particular risks.

Each of these works in a different way to convince people to avoid exposing themselves to the contamination.

In some states, risk assessment plays a prominent role in decision-making at hazardous waste and other contaminated sites. For example, many Northeastern states in the United States support well-defined programs for site investigation and risk-based decision-making. Specifically, the Massachusetts Contingency Plan (MCP) stipulates a three-tiered risk assessment approach based roughly on the tiered approach originally developed by ASTM for the evaluation of petroleum-contaminated sites (http://www.state.ma.us/dep; see Section III). The MCP contains policy decisions that depart from those in place at the federal (USEPA) level, specifically with respect to the manner in which site contaminant concentrations are determined (arithmetic average concentrations as directed by the MCP versus the 95th percent upper concentration limit on the arithmetic mean at the federal level) and the acceptable level of carcinogenic risk used to determine, for example, cleanup concentrations (10^{-5} as per the MCP; a risk range of 10^{-4} to 10^{-6} as per the USEPA). Risk assessment at the California USEPA and the State's Department of Toxic Substances Control is conducted using state-specific guidance concerning default exposure parameters, toxicity factors, and mathematical models for determining total chemical exposure (http://www.dtsc.ca.gov). In contrast, risk assessment guidance in Southeastern states, such as Florida, North Carolina, and South Carolina, at best relies heavily on existing USEPA guidance, at worst is nonexistent, or only makes vague references to "risk-based" principles.

XII. INTERNATIONAL RISK ASSESSMENT PERSPECTIVE

A recent publication in the journal *Land Contamination & Reclamation* (Ferguson, 1999) summarizes risk assessment policy and practice in 16 European countries (Austria, Belgium, Denmark, Finland, France, Germany, Greece, Ireland, Italy, The Netherlands, Norway, Portugal, Spain, Sweden, Switzerland, and the United Kingdom). The countries are participating in the Concerted Action on Risk Assessment for Contaminated Sites (CARACAS) initiative of the Common Forum for Contaminated Land of the European Union. This publication provides international perspective to risk assessment.

In a number of countries—Austria, France, Greece, Ireland, and Portugal—no formalized process for the assessment of contaminated sites existed as of the date this journal article was published.

- In Austria, there are no general "intervention values" for the evaluation of polluted soils. It is preferred to base evaluations on site-specific circumstances, especially local geologic conditions and anthropogenic influences on soil quality. Since more than 99% of Austria's drinking water is supplied by groundwater, there is a very strong emphasis on the prevention of groundwater pollution. In this regard, the Austrian Water Act is characterized by its use of the precautionary principle. The precautionary principle involves decisions about the best ways to manage or reduce risks that reflect a preference for avoiding unnecessary health risks instead of unnecessary economic expenditures when information about such potential risks is unavailable or incomplete (see discussion in Chapter 2). In common terms, this position is similar to the "better safe than sorry" approach.

- France has no specific legislation concerning contaminated sites. Technical guidance for detailed risk assessment at such sites is still under discussion.

- In Greece, no national guidance documents on risk assessment currently exist for contaminated sites. Guidance documents have been developed by some organizations, but they do not have any general enforcement authority.

- Ireland also lacks specific contaminated land legislation, although a National Hazardous Waste Management Plan is currently under preparation, and will establish a framework for the management of sites that have been used in the past for disposal of hazardous waste.

- Portugal has not yet compiled data on contaminated sites, nor has it established national methodologies, criteria, or explicit risk procedures for their assessment and remediation.

In Belgium, soil cleanup values have been established and are defined as levels of soil pollution above which serious harmful effects for man or the environment might occur. An exposure assessment model has been used to derive the soil cleanup values. It is based on formulae used in a Dutch model (described in the following paragraph). For each pollutant, exposure calculations for relevant exposure scenarios were undertaken to estimate a total exposure equal to the Tolerable Daily Intake (TDI) for non-cancer effects, or the dose corresponding to a theoretical excess lifetime cancer risk of 10^{-5}. Values for TDI and unit cancer risk are taken from international (*e.g.*, World Health Organization, USEPA) databases.

In Denmark, risk assessment is based on determining contaminant concentrations and comparing them with quality criteria for soil, groundwater, and air. In September 1998, a new Guideline on remediation of Contaminated Sites was issued. Under this guideline, risks are assessed for sensitive lands such as housing with gardens and children's playgrounds. Topsoil quality criteria for approximately

50 substances have been developed based on human toxicity. The decision receptor is usually taken to be a two-year-old child who is assumed to eat 0.2 gram of soil per day, or on isolated occasions, 10 grams of soil. Groundwater quality criteria have also been derived for approximately 50 substances. Assessing the risks from volatile soil contamination in relation to indoor air is based on contaminant transport by diffusion through pore spaces in the unsaturated soil zone and transport by convection into buildings through gaps in concrete floors, similar to the RBCA model utilized in the United States.

In Finland, measurement-based assessment of risk at contaminated sites is usually done by comparing observed concentrations with guideline values for soil contaminants. Preliminary guideline values for approximately 170 substances have been developed, based mainly on Dutch values. A two-level guideline system includes target and intervention concentrations derived mainly based on ecotoxicity but also including some human health considerations. Values for different land uses have not been presented due to emphasis on long-term multifunctionality of soil.

In Germany, risk assessment is understood to mean the whole process of site evaluation following an initial historical investigation. Risk assessment is carried out case-by-case and decisions depend on the type of land use, the degree and extent of pollution, the relevant receptors, and the existence of exposure pathways. The Federal Soil Conservation Act was ratified in 1998 and came into force in 1999. The Draft Ordinance on Soil Conservation and Existing Contaminated Sites contains the following proposed risk-based values as:

- soil screening values for the direct soil-to-human pathway for different land uses: children's playgrounds, residential, parks and recreation, industrial and commercial;

- soil screening values for the soil-to-edible-plant pathway;

- leachate screening values for the soil-to-groundwater pathway; and

- action values for the direct soil-to-human pathway.

Screening and action levels are both risk-based. They are based on simplified exposure scenarios, such as soil ingestion for children playing outdoors, rather than on all theoretically possible exposure pathways. Soil screening values are also based on:

- A set of toxicological reference dose (TRD) levels which give a virtually safe dose via ingestion or inhalation, and based on this virtually safe dose, an estimation of a body dose that indicates a certain level of risk to public health;

- Exposure from ingestion or inhalation of soil based approximately on the 95th percentile intake for an exposed population; and

- Substance-specific considerations—e.g., bioavailability and background concentrations.

Effects resulting from combinations of substances have not been considered thus far. For carcinogens, as a starting point for deriving trigger levels, a theoretical excess lifetime cancer risk of 5 x 10^{-5} is suggested for each individual substance.

The main guidance documents for conducting site-specific risk assessments in Italy have been the USEPA's *Risk Assessment Guidance for Superfund* and ASTM's *Risk-Based Corrective Action*. That is, risks to human health are evaluated through exposure assessment of target populations or individuals, both for present and future exposure. Dose-response assessment (toxicity and carcinogenicity) is integrated with exposure assessment to provide a quantification and characterization of risks. Cleanup objectives and remediation goals are set using results of the three-tiered RBCA procedure.

Risk-based soil quality objectives are an important instrument in Dutch soil policy. Target values and intervention values have been established for about 100 substances for soil and groundwater. Exceeding such values indicates the potential for risk, assuming that exposure occurs to its full extent. Dutch guidance recognizes, however, that full exposure will not always occur, and in these instances, "local circumstances" are taken into account when estimating actual risks. In the Netherlands, local authorities, provinces, and municipalities are largely responsible for the use of instruments such as soil quality objectives and risk assessment procedures.

In Norway, the most important provisions concerning pollution of the environment are gathered into one law, the Pollution Control Act of 1981. Under this law, a two-tiered decision model was developed. In the first tier, generic target values were developed for several metals, volatile chemicals, PAHs, PCBs, and mineral oil. These target values, which relate only to the most vulnerable land use, are based on existing Dutch and Danish values for contaminated sites. For other land uses, or for occasions when target values are exceeded, a second tier – the site-specific risk assessment – is applied.

A national inventory of contaminated sites is being developed in Spain using Dutch guidance values. Of the sites examined, 4,900 have been found to be "potentially contaminated" and 390 of these have been investigated in detail using a risk assessment matrix approach that evaluates contaminant toxicity, mobility, and risk to potential receptors. In the Basque Country, soil quality is defined based on risk assessment for "protected targets" (human health and the environment) and intended land uses. Soil screening values are known as "Indicative Values for Assessment." These values have been developed and are land-use dependent and provide a generic assessment that will allow essentially risk-free soils to be differentiated from soils that pose or could potentially pose risks for the intended use.

In Sweden, risk assessment involves identifying and describing the risk of adverse effects on human health and the environment at contaminated sites. The assessment is based on high-end (but not implausible) exposure scenarios. The Swedish Environmental Protection Agency has developed guideline values for 36 contaminants or contaminant groups in soil. For each substance, guideline values have been developed for two different types of land use:

- Land with sensitive use (*e.g.*, Residential areas, kindergartens, agriculture); and

- Land with less sensitive use (*e.g.*, Offices, industries, roads, parking lots).

Generic values have been derived using a Swedish exposure model based on similar models and data developed by other countries and international organizations.

In Switzerland, the identification, assessment, remediation, and financing of contaminated sites are regulated by the Federal Environment Protection Law. According to current estimates, about 50,000 polluted sites exist within the borders of the country. About 3,500 of these may require some degree of remediation. In order to identify the small number of "dangerously" polluted sites within the large number of contaminated sites, a site-specific risk analysis based on interactions between the site and the environment—mainly groundwater, surface water, soil, and air—is required. The site-specific analysis takes into account the potential for transport and barriers. Intervention values for landfill leachate and air are derived based on human toxicity.

The Environmental Protection Act of 1990 provides a regime for the control of specific threats to health and the environment from existing land contamination in the United Kingdom. Contaminated land is identified based on risk assessment. "Precautionary threshold trigger values" are used as screening levels for some of the more common soil contaminants. Detailed, sites-specific risk assessments, based on exposure and toxicity assessments, are used where these trigger values are not available, not appropriate, or where particularly complex or sensitive site circumstances require it. In the context of direct human health risks, these trigger values are being replaced by guideline values. These values are derived by employing the same procedures and algorithms used in detailed site-specific risk assessments, but applied to typical land use scenarios characterized by specific exposure assumptions. "RBCA-type" risk assessments—*i.e.*, simple screening approaches followed by more detailed and sophisticated risk assessment methods, if justified—are used to evaluate groundwater contamination. In some circumstances, it is necessary to consider harm to or interference with ecosystems and habitat.

Appendix B

EVALUATING FINANCIAL LIABILITY IMPLICATIONS OF ENVIRONMENTAL RISKS

John Rosengard

I. A FINANCIAL PERSPECTIVE

Most corporate environmental programs have active and successful methods for addressing threats to human health and the environment. Once immediate threats are under control, long-term regulatory compliance and shareholder values require balancing. At the extremes, a company may selectively comply with environmental regulations or make environmental capital expenditures beyond any business justification. Either extreme results in failure. Each major company today operates between those extremes, while working in response to its shareholder's values.

Two principal shareholder values are the survival of the corporation and a competitive return on capital employed. All other values, such as corporate citizenship, for better or worse, are less important. The measurable consequences of environmental spending (such as cost, waste reduction, and energy conservation) deserve periodic analysis from the shareholder perspective:

1. Does environmental spending—whether capitalized, expensed, or accrued—add value to the corporation?

2. Do existing waste management practices pass a periodic financial risk analysis?

3. In adding value, does the spending reduce liabilities, reduce uncertainty around those liabilities, or both?

Working in the best interest of shareholders is one of the most responsible things companies do. This makes a company's actions predictable and rational. Every decision allocates the company's capital to achieve some benefit at some cost. Therefore, there is a continuous need to:

- scan for contingent environmental liabilities, including expenses, investments, and potential income;

- benchmark with peer companies;

- apply financial risk management techniques;

- build consensus for effective environmental strategies across corporate functions (especially regarding remediation of impaired property); and

- train corporate environmental project managers to factor business issues into their work, that is, to quantify shareholder wealth into their decision.

Through developing shared understandings among a team of people (from operations to finance, legal, and real estate), intuitive decisions emerge from this consensus building, along with "like-an-owner"-thinking that incorporates financial risk and uncertainty. These decisions build a dialogue on the many corporate issues (identified in Chapter 2), critical assumptions (made in scientific, engineering, financial, and management analyses and decisions), and alternative solutions, and thereby translate into lessons for the entire portfolio.

Two visible metrics for environmental project teams are meeting annual budgets and preventing sudden and major environmental reserve increases. In the mid-1990s, one United States energy company stated its environmental reserves at more than \$500 million, but further disclosed that the actual costs may be more than double the reserved amounts. For this company, the financial loss exposure to shareholders meant that six months of corporate income were now set aside and that another six months might be needed. To shareholders, this means a full year of profits is needed to pay for one group of environmental projects.

As corporate managers review their decisions in the context of the company's survival and return on capital, environmental contractors (engineers and consultants) should also begin measuring their proffered and designed solutions differently. They should offer these solutions with an understanding not simply in terms of a near-term budget, but in terms of lifecycle costs and foregone opportunities to close a project sooner, or at a more narrow range of costs. With environmental remediation liabilities, for example, success should be redefined as lowered contingent liability, instead of tasks completed "on-time, on-budget."

Incorporating effective risk management into portfolio disclosure is not only useful for discharging contingent liabilities from past and current operations, but can be a source of advantage for growing companies. If an acquired business has existing environmental remediation reserves, environmental operating expenses, and upcoming environmental capital projects, the acquirer can add value to each type of spending in different ways:

- For the environmental reserves, in an acquisition, there is a stated book value to the liabilities. The acquirer adds value by extinguishing those liabilities for less than book value, and loses value if the reserves must increase later.
- For environmental operating expenses, the acquirer creates value by cutting these repeating costs.
- For capital expenditures, the acquirer may add value through a lower cost of capital, sharing or deferring the initial cost, using increased capital expenditures to replace operating expenses, and by accelerating the benefits.

As mergers and acquisitions occur, there are variations in the values placed on a variety of assets and liabilities by buyer and seller. Environmental liabilities are not that different.

II. FINANCING ENVIRONMENTAL RISK MANAGEMENT

Corporate environmental risks use one of three types of funding, each with its own specific uses, tax consequences, and profit implications.

A. CAPITAL EXPENDITURES

An environmental project, when treated as a capital expenditure, has the following attributes:

- a useful life span;
- some contribution to future profits; and
- a significant initial cost.

A capital expenditure has an initial cash outlay, but a deferred expense effect, depreciation, over the useful life span of a project, which generally runs from three to thirty years. For example, when a plant adds an air scrubber, which should last ten years, the purchaser pays the cost at the beginning of the useful life, but, for tax purposes, reports the scrubber's cost on its income statement and tax returns at the rate of 10% per year for ten years, using a "straight-line" method of depreciation. While there are at least ten different depreciation methods, the general purpose of this type of funding is to roughly match the expense and benefits of an investment, on an annual basis.

B. OPERATING EXPENSES

Most routine and small environmental costs are expensed, meaning the cash cost is treated as a business expense the moment it occurs. The electrical cost to run an air scrubber, for example, would normally be expensed on a running basis.

C. RESERVE EXPENDITURES

The establishment of environmental remediation reserves occurs when a company identifies and expects future cleanup costs for a given project. These projects are usually at divested or discontinued operations or at multiparty sites (*e.g.,* Superfund sites). When a company determines that a cost is both probable and reasonably estimable, a reserve (or liability account) is created or an existing one increased. Additionally, an identical operating expense for those costs is declared that year, although actual expenses may not be paid for several years; and may increase or decrease. For tax purposes, this is not a restatement of prior years' earnings, but an expense against current profits when the reserve balance is created or increased. The justification for establishing reserves is to recognize changes in a company's overall financial position at that moment when sufficient information is known, and to separate future earnings from remediation expenses for discontinued businesses or waste management practices.

D. COMPARISON OF FINANCING TYPES

The following table illustrates the differences between the three funding types described above.

	When cash is spent	When expense is declared
Capital Expenditure	Now	Future (useful life)
Operating Expense	Now	Now
Reserve Expenditure	Now	Past

Table B.1 presents the effects different funding mechanisms for environmental projects have on profitability and shareholder wealth.

Environmental remediation liabilities, depending on the company and the information known about a site, are established for spending anywhere from one to

Table B.1

Funding, Profitability, and Stakeholder Wealth

	Effect on Profitability (Income Statement)	Effect on Shareholder Wealth (Equity)
Capital Expenditure	Continuous expense over entire project life cycle	Usually positive, especially for discretionary projects
Operating Expense	Negative	Negative; continuous pressure to defer if simultaneous benefits are not tangible
Increasing Environmental Reserve	Negative	Larger liabilities, smaller shareholder equity
Spending Environmental Reserve	None	Neutral; decrease in cash (asset), decrease in liabilities

forty years into the future. For example, there may be several decades of monitoring expenses at a formerly used mine or landfill. Alternatively, there may be a one-time "cash out" payment to settle a company's Superfund liability at a multiparty site.

III. BUILDING SHAREHOLDER VALUE FROM PROJECT COSTS

Building a life cycle cost for an environmental project requires analysis of consequential costs and income. For example:

- Removal of an underground storage tank at a gasoline service station— many States have reimbursement programs, where remediation and tank replacement costs may be paid from taxes collected at the pump. This reimbursement is *taxable* income to the company.
- A Brownfields remediation project at a surplus (vacant) facility—through cleanup, an unmarketable property may become more valuable. The difference between the book value of a property and the selling price is a capital gain to the corporation.
- Real estate operating costs—if a surplus facility is fenced and inactive the owner remains responsible for property taxes and site security and will still take depreciation expenses for property improvements.
- Opportunity costs—when an inactive property is owned by a company (at the property's "book value") but no income is generated by it. If the company has a $1 million property and a 10% return on capital target, factoring in an annual opportunity cost of $100,000 may alter the life cycle cost estimates to favor one strategy over another.

The purpose of these examples is to demonstrate that environmental projects can generate consequential income and expenses, and that net life cycle costs, after taxes, accurately represent the impact to the shareholders.

Two other factors that more fully describe shareholder impact are cost uncertainty and timing:

- Cost uncertainty is the restatement of a dollar cost point estimate as a range, and statistically correlating those ranges so that the net life cycle cost is defensible and reproducible.

- To account for timing, corporations evaluate future revenue and expenses with a filter called discounting. Discounting is where an expense, that occurs this year, is valued differently than an identical expenditure in a future year. This is done for three reasons:

  o First, prices generally increase year after year. For example, an inflation rate of 3% states the general expectation that a contracted cost for $1,000 this year should be about $1,030 next year, $1,061 the year after, and so on.

  o Second, the company has return-on-investment (ROI) goals, based on its cost of capital and the opportunities in its principal lines of business. Using a recent prime lending rate of 8.25%, and a 3.0% expected inflation rate, the company's "return on investment" target would be the *product* of these two values:

  $$11.5\% = (1+0.0825)\times(1+0.03)-1$$

  Terms like "return on capital employed," "economic value added," and the "internal rates of return" derive from a company's ROI goals.

  o Finally, the company must consider investment-specific risks. This is because inflation rates and ROI targets are not fixed over the long-term, but fluctuate due to other factors, such as the fact that a longer-term project has exposure to more *potential* cost-influencing factors than a shorter-term project.

The following equation shows how to calculate the discount rate:

$$\text{Discount Factor} = \frac{1}{[1+(\text{ROI Target})]^n}$$

Where: n = number of years from present. For example, if a company estimates project costs at $50,000 in two years, the discount rate would be 0.8044.

$$\text{Discount Factor} = \frac{1}{[1+(0.115\pm0.00)]^2}$$

$$\text{Discount Factor} = \frac{1}{1.24} = 0.8044$$

The resulting present value would be the product of $50,000 times 0.8044 or $40,220. If the same company has a $200,000 project with costs spread out over four years, using a discount rate of 11.5%, the resulting present value is $153,481. The following calculation demonstrates the discount rate and present value calculation for a groundwater remediation project:

	Year 1	Year 2	Year 3	Year 4	Present Value
Groundwater Pump/Treat	−$50,000	−$50,000	−$50,000	−$50,000	−$153,481
Discount Factor	1.0000	0.8969	0.8044	0.7214	

IV. COMPARING AND OPTIMIZING STRATEGIES

Discounting gives project managers a basis for financially comparing the present values of several multi-year projects. Below is a comparison of the present values of three different environmental remedy strategies. While each strategy has a cash cost of $200,000 over four years, their present values range from –$153,481 to –$179,372. The range in value is due to the differences in the timing of the expenditures. The present values are negative because each alternative *subtracts* from the current value of the corporation. From a purely financial perspective, the groundwater pump/treat remedy is the best choice because of the financial objective of <u>minimizing</u> negative present values.

	Year 1	Year 2	Year 3	Year 4	Present Value
Groundwater Pump/Treat	–$50,000	–$50,000	–$50,000	–$50,000	–$153,481
Soil Vapor Extraction	–$100,000	–$33,333	–$33,333	–$33,333	–$162,110
Excavation	$0	–$200,000	$0	$0	–$179,372
Discount Factor	1.0000	0.8969	0.8044	0.7214	

Figure B.1 shows what happens when point estimates of net present values are converted to ranges. This kind of analysis greatly increases the information available for making a decision about competing environmental strategies. Using the present value point estimates in the table above, project managers have no information about which technologies have wide cost ranges (say for example $100,000 to $500,000) versus narrow cost ranges (*e.g.*, $150,000 to $250,000). For risk-averse companies or situations (such as likely litigation or high-visibility projects), comparing pessimistic cost estimates may be appropriate.

V. EVALUATING A CORPORATE PORTFOLIO

Aggregating life cycle cost estimates for environmental projects requires some financial sophistication to generate accurate cost ranges. First, costs and income need examination from a tax perspective to separate capital, operating, and reserve expenditures. Second, the future expenditures against environmental reserves must be divided into costs that are both *probable* and *reasonably estimable*, and those that do not meet both criteria. Costs that meet both criteria must be reserved, while other costs may be reserved in the future, when the costs are more probable and/or reasonably estimable.

Under evolving public disclosure requirements, United States companies will collect more information about environmental liabilities for annual reports to shareholders. It also is likely that many companies, by pursuing existing practices, will continue to increase their environmental reserves periodically. As these reserves are spent, two things happen: sites are remediated and, to some degree, reserves go back up. Superfund and RCRA expenses are typically a very small percentage of the revenue of a *Fortune 500* company. However, taking a charge

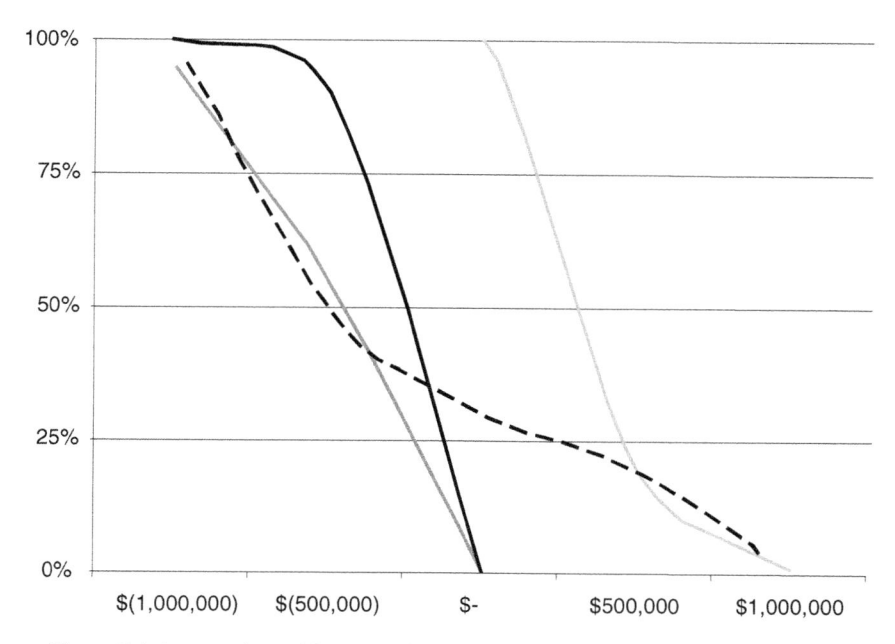

Figure B.1 A comparison of four scenarios using reverse cumulative probability distribution function.

(that is, increasing the level of reserves) for five to ten years of estimated expenses can be a substantial portion of a year's net income. Preventing surprises to shareholders, in the form of sudden and major reserve increases, can be achieved by regularly calculating expected environmental reserve, operating and capital expenditures for discontinued and operating facilities, and addressing the degrees of risk and uncertainty from the budget line item level up through the entire portfolio of contingent liabilities.

The pyramid diagram in Figure B.2 depicts the flow of risk-ranged assumptions throughout the environmental project cost estimating process. Once a manager can bracket the estimate of a budget line item (at the bottom of the pyramid) to reflect current experience and expectations, a single site probability distribution is calculated (middle of the pyramid). Next, the various single site ranges (moving up the pyramid) are used with the ranges of other sites to calculate the value of the groups of projects (sub-portfolios) and the entire corporate portfolio (top).

In placing a value on an environmental portfolio, it is essential to adjust one's perspective continually, while complying with the accounting and tax requirements. While it requires diligence to separate operating, reserve, and capital expenditures, buyers and sellers differ fundamentally in how they value any liability due to their cost of capital, target return on investment, tax rates, disclosure procedures, reputation with environmental regulators, and long-term plans for an ongoing business.

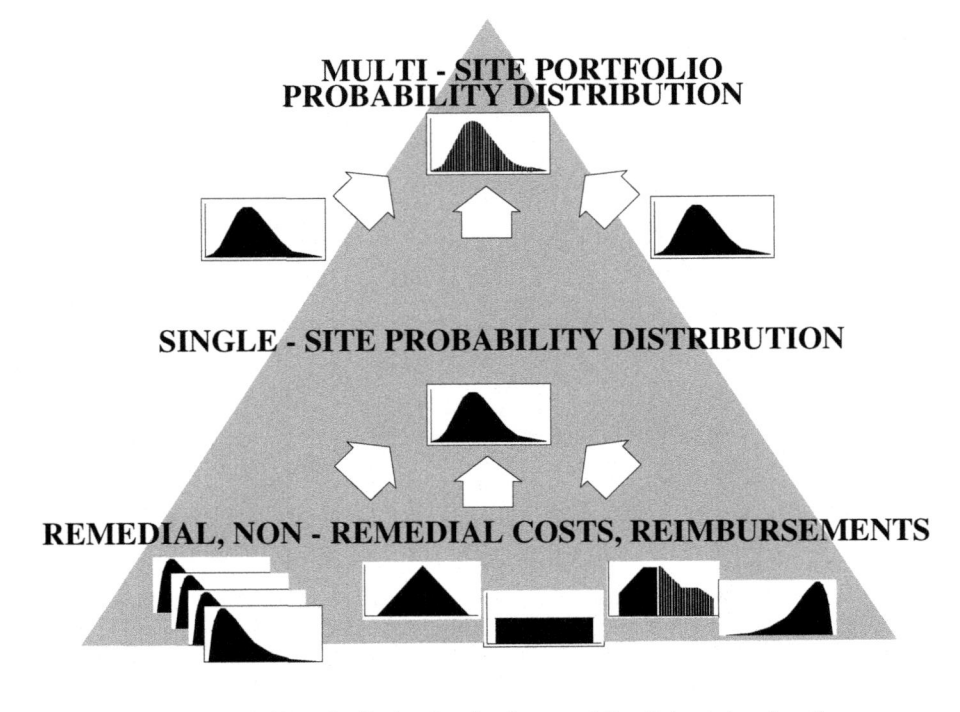

Figure B.2 The probability distribution function for a portfolio of sites is based on the range for individual sites, and in turn, the ranges for study, remediation, and related costs.

Capital gains from the sale of an environmentally impaired property *should* be part of the remedy selection process, but *is not* part of the environmental liability portfolio reserve calculation. Similarly, reimbursements, depreciation, and insurer recoveries are *not* part of the environmental remediation liability reserves. Operating expenses, capital expenditures, and income, while sometimes vital to defending a strategy selection decision at a single site, are reported by corporations separately from the remediation reserves.

Turning back to the reserve calculation, once a project has costs that are probable and reasonably estimable, accounting guidance allows a subsequent step to determine the proper reserve estimate for a site. If a site has a best estimate, that number is the recommended reserve, but if there is no better estimate within a range, the lower end of that range may be used as the recommended reserve for that site.

Several statistical challenges exist in generating and assessing the cost ranges for each reserveable site:

1. An estimator may expect that the sum of all optimistic budget items for a project will equal the optimistic cost estimate for the entire project or site. This presumes a perfect statistical correlation of all cost assumptions. For higher-cost sites, where uncertainties exist around soil volumes, target contaminant concentration levels, and duration of post-closure monitoring phases, there may be no defensible reason to correlate the cost assumptions.

2. There will be inadequate data. Generating cost ranges based on experience, contractor bids, Records of Decision, and cost databases is difficult at best.

3. Environmental remediation projects tend to have very little in common, so translating spending lessons to other sites is complex.

4. There are fundamental barriers to the productive exchange of cost information. Contractor turnover, attorney-client privilege, lack of resources and tools, and regulator flexibility create circumstances where similar sites can be estimated in different ways.

Modern portfolio theory tells us that with a sufficiently large number of sites, the extreme high and low cost outcomes on a few sites will cancel each other out. This will result in an expected value, for the portfolio, that is roughly the sum of the expected values for each site.

Table B.2 and Figure B.3 present a set of reasonably estimable environmental expenses (a subset of undiscounted project life cycle costs) for each site to support the summation of low and high estimates. Table B.2 and Figure B.3 also present a statistical model of the full portfolio in order to create a reserve forecast. The following describes the model, its features, and findings.

- Table B.2 indicates that the range of risk in this portfolio is $10.1 million to $40.7 million.
- If the "Low" and High" columns are simply totaled, we would expect the total required accrual to be in the range $10.1 million to $40.7 million, with an expected value at the peak value shown.
- Using a Monte Carlo portfolio simulation analysis, where the low, mid, and high values are assigned probabilities, we see the second, smoother line, which parallels what portfolio theory tells us: reaching the sum of the "low cases" across a large portfolio is impossible.
- The true range is much narrower (*i.e.*, $12.5 million to $35.0 million) than the range created by summing the low and high estimates (*i.e.*, $10 million to $40.7 million). Moreover, the midpoint of the correlated range is lower than the simple calculation using a triangular sum approach.

Building the reserve calculation process from the site-specific budget line item up to the entire portfolio is vital. This is because companies have the option of reserving the low end of a range, instead of the midpoint, and the methodology must meet American Institute of CPAs auditing guidelines on environmental remediation liabilities (SOP 96-1 October 1996).

Given the nature of environmental projects, and the flexible criteria of "reasonably estimable" cost recognition, one of the few certainties companies have is to work in probabilities. Having and using consistent definitions (*e.g.*, likely = p.10 = 10th percentile = 10/90 estimate) and modeling them properly is the best approach to uncovering a site's true financial risks and the portfolio's financial uncertainties to the company caused by environmental risks.

Table B.2
Liability Estimates for Reserve Forecasting Purposes

| Project | COST RANGE | | |
	Low $	Mid $	High $
1	27,000	65,000	142,000
2	36,000	69,000	135,000
3	47,000	90,000	176,000
4	28,000	82,000	189,000
5	44,000	130,000	297,000
6	12,000	19,000	27,000
7	14,000	27,000	53,000
8	93,000	278,000	628,000
9	65,000	126,000	244,000
10	85,000	134,000	191,000
11	211,000	513,000	1,108,000
12	202,000	394,000	758,000
13	103,000	306,000	695,000
14	137,000	324,000	719,000
15	172,000	331,000	645,000
16	275,000	433,000	619,000
17	290,000	698,000	1,523,000
18	560,000	1,670,000	3,780,000
19	462,000	1,139,000	2,426,000
20	897,000	1,731,000	3,364,000
21	903,000	1,755,000	3,386,000
22	861,000	1,355,000	1,937,000
23	629,000	1,941,000	4,246,000
24	941,000	2,248,000	4,940,000
25	980,000	1,543,000	2,205,000
26	630,000	990,000	1,418,000
27	819,000	1,287,000	1,843,000
28	576,000	1,394,000	3,024,000
Portfolio	10,099,000	21,072,000	40,718,000

Best Estimate = Reserve Value

Note: This listing of 28 projects shows some sites where the lower estimate is appropriate for reserve forecasting purposes, and some projects where the mid or high cost estimates are the estimator's best estimate of the liability.

VI. THE NEED FOR BETTER ANALYSIS

Several aspects of corporate environmental project management highlight the need for better business analysis of the options available to the company; these include:

- treating project completion as the endpoint, over and above stakeholder and regulator satisfaction, or limiting exposure to current and future risks;
- excluding life cycle costs, reimbursements, capital expenditures, and operating business impacts from decision analysis;
- ignoring land use, property taxes, and property value from decision analysis;

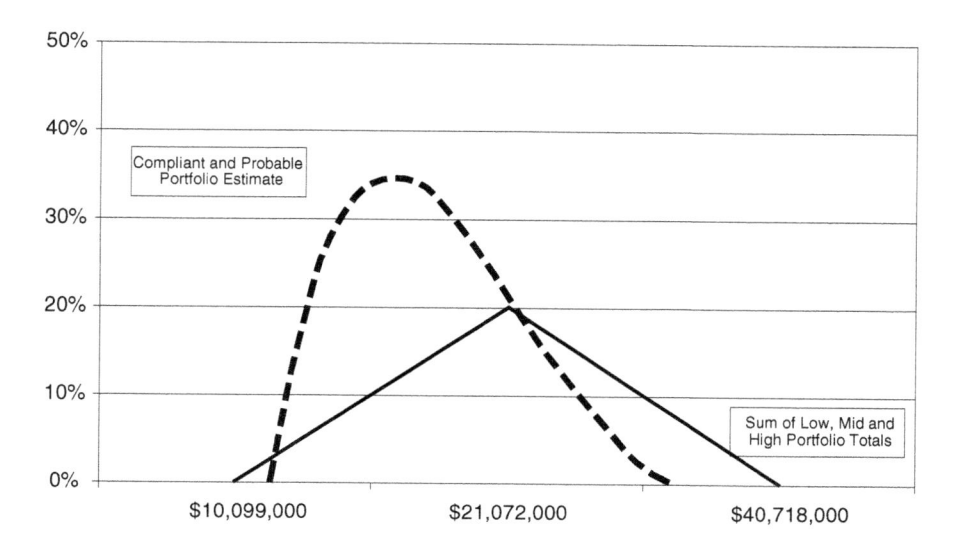

Figure B.3 This figure shows the potential for misinterpreting a liability portfolio when aggregating low, mid, and high estimates for projects. A more likely outcome, the smoothed curve, reflects the likelihood of conservative and optimistic estimates canceling out one another.

- being out-negotiated by contractors, regulators, insurers, and other potentially responsible parties;
- inconsistent strategies and costs for similar projects; and
- poor understanding of how spending narrows cost ranges, reduces liabilities, or both.

Environmental liabilities are one of three major contingent liability types facing larger companies today. The others are product warranty/liability costs and post-retiree medical costs. The accounting profession has similar tests to check if a company has adequate reserves for each liability type, but there is no standard method for calculating the reserves for any contingent liabilities.

It may not be possible to realize or calculate every perceived environmental risk in financial terms today (for example, the impact of a company's products on biodiversity or global warming). Nevertheless, there are several statutory requirements for converting environmental risks into financial risks. Environmental risks are sequenced into financial risks through continuous screening and funneling of data to corporate management, and if material, to shareholders.

At a minimum, environmental liability management has to focus on material liabilities for the corporation:

- how those liabilities occurred and how they are being discharged, and
- what is the known current and the potential future cost.

Each company has a different materiality threshold, in part because the definition is flexible, and based on a company's performance. Working definitions for materiality are generally at least 3 to 10% of total liabilities, or reserve increases that are at least 3 to 10% of corporate operating income.

When a liability is material, generally accepted accounting practices and SEC filing requirements obligate public companies to accrue and disclose the liabilities, and display any material known information. Whether or not environmental liabilities are material, a greater level of effort is needed for estimating, monitoring, and reporting environmental liabilities internally. It is at this level where financial risk analysis has a significant impact.

Appendix C

RISK COMMUNICATION BASICS

Samuel D. Ostrow

I. INTRODUCTION

Assessment of a polluted site is science. Management of a site and the remediation of its pollutants is science. Communicating about the past and potential impact of the site on the physical, social, and economic health of the adjacent community is art.

Polling, focus groups, and the other measurement techniques usually associated with science guide communications experts. Yet, ultimately, when you inform a community that a contaminated site is in their midst—and that it needs remediation to protect health and the environment—you deal with powerful and unprogrammable emotions and a complex of interest group interactions that move deliberation from scientific probabilities to the anticipations of the artist.

II. "THE RULES"

Fortunately, there are models for anticipating public behavior in any situation in which a public learns that it has been exposed to a risk—any risk. Early in the 20th century, the German physicist Werner Heisenberg announced what has become known as the Uncertainty Principle: the mere act of measuring an object changes it. Being associated with an environmentally impaired site has a similar impact on how a company is perceived by its different publics. Whatever its reputation before, it is now inherently and inevitably changed. This model leads us to three rules of managing the public issues involved with site remediation.

RULE 1. Once brought into the public eye because of your association with a toxic site, forget about winning and losing. Focus only on restoring that sense of trust with each of your constituencies on which your relationship with them begins. It is of great benefit to remember two simple facts:

1. In virtually every issue's controversy, it can be safely assumed that about 10% of the public hated you (for whatever reason) before the controversy, hates you now, and will continue to hate you no matter how good a job you do in resolving the situation.

2. Another 10% of the public loves you, and they will continue to love you regardless of how toxic the site is, no matter how deformed the seventh generation litter of puppies.

RULE 2. The "facts of life" of Rule 1 lead to the second rule of communications about environmentally impaired property: it is a waste of time to argue with either the lovers or the haters. It is not possible to change the haters' minds, and nothing of value will be obtained from preaching to the already faithful. Instead, focus on that 80% of the public who do not ordinarily think about either you or the issue. They will make up their minds in the 90 seconds that television gives to the story. Focus also on the 10 seconds within the 90 seconds that will be given to your sound bite. Within that brief time, each of your constituencies will determine whether you or the opposition is being more reasonable in addressing their concerns. If those constituencies select you, recognize that they have done little more than give grudging acceptance to what it is you have proposed.

RULE 3. The third rule is the hardest for most engineers and executives to accept. You may not think your contaminated site is a crisis. Your staff and other internal and external stakeholders may not think it's a crisis. However, if the media says, "it's a crisis," then it's a crisis; and you will be held accountable to the rules of behavior and communications the media has established for "companies-in-crisis." Those rules have never been written down—that way, it is easier for reporters to change them—but there is a reasonably persistent pattern of reporting on environmental crises, and the media have trained their audiences to expect this pattern and evaluate corporate performance against it.

III.   THE FOUR PRIORITIES OF PUBLIC COMMUNICATION

There are four essential priorities to address when implementing risk communications pertaining to environmentally impaired properties.

A.   FOR EACH AUDIENCE, DEFINE THE RISK
Describe and then provide a sense of measurability to the risk or risks of greatest concern to each. The risks will differ from audience to audience, although the communication is related to the same site:

- For the neighbors, the risk of greatest concern may be the risk to health.

- For investors, the risk is the company's potential liability.

- For customers, it is whether the costs may go into rates.

- For government officials and regulators, the risk may be whether they can be seen by their constituencies as exercising less than adequate oversight.

Where the magnitude of risk is unknown, experience indicates that the wiser course is to initially define the risk as potentially large. This impression creates the sense of the risk potential being managed as the actual magnitude is narrowed and determined more accurately.

B.   DESCRIBE ACTIONS THAT MITIGATE THE RISK
What will the company do? What will third parties and government do? What must individuals themselves do? Early on, after the need to remediate a site has been determined, the description of mitigation may not and frequently cannot be the company's full remediation plan. The focus here, fortunately, is intensely short-term. You may want to tell neighbors to avoid the site (or that there is no need for them to do anything). You may want to tell them that there is a particular type of

screening they should undergo (or that they need not do anything because there are no expected consequences). You may want to tell them that the company is developing a cleanup program that will remove or reduce whatever level of risk exists. You also may want to tell them that there are government programs that will act to mitigate the risk. Again, the key is not so much providing a detailed plan, as it is the sense that the risk can be mitigated.

C. IDENTIFY THE CAUSE OF THE RISK

How was this site polluted? There are two reasons why this priority is a critical communications element:

- First, the public has been trained from reports of airplane crashes and similar disasters to assume that if you know what went wrong you know how to fix it. That is why we fly on airplanes and ride trains. We actually believe that they will change the part or procedure that led to the previous crash. We know the pollution of the type associated with MGP plants, and we certainly aren't going to build any more of them, at least without the protection that contains and then disposes of their effluents safely.

- The second reason to focus on cause identification is that it shows you understand the media game. Mysteries and crime-solving shows remain among the most popular television programs because everyone wants to know "whodunit?" If you are part of the process of helping media report, "whodunit," you demonstrate an understanding of their imperatives that will stand you in good stead throughout their coverage of the particular issue. This activity is not accomplished without pain, to be sure, but experience strongly suggests that, to the majority of the public, when a company blames a system or a process within its control, it is considered a sufficient admission to restore trust.

D. DEMONSTRATE RESPONSIBLE MANAGEMENT ACTION

More than anything, this priority comes down to communicating and acting in a manner that says it is the company that is in charge of resolving the issue—not the media trying to turn each related event into a vehicle for stimulating ratings, and not self-serving politicians seeking to find a political advantage. Part of demonstrating responsible management action clearly has to do with how well the company communicates the other three priorities. However, it also includes:

- communicating the clear sense that the company is acting according to a plan;

- that it has a situation management strategy in place;

- that it is working proactively and productively with the appropriate authorities;

- that its senior executives are appropriately engaged; and

- that it has available executives at appropriate levels of the company's functional and hierarchical levels who are able to communicate the company's messages coherently and professionally, again, to the standards the media will impose in situations such as these.

IV.   DECISION PRIORITIES

Knowing the four priorities of communicating about issues is not the same thing as knowing what to say. The decisions about what to say also cannot be left only to the communicators, only to the environmental engineers, or even only to senior management. A company that knows it will be confronting a series of issues over the usual long-term process of environmental cleanup needs to put in place an integrated team to manage and make decisions about the communications requirements. That team should be guided by a series of decision priorities that assure that each act of communication is based on an efficient consideration of all issues that impact what a company says and how it says it.

A typical set of decision attributes used in managing a remediation might include the following elements.

A.   TO WHOM MUST WE COMMUNICATE AND IN WHAT PRIORITY?

This element is not as simple as it sounds. The answer to the question may vary depending on the particular development about which the company is communicating. In some cases, the highest priority audience may be the site's neighbors. In other cases, it may be the regulatory authority with the greatest interest in the cleanup. It also may be local politicians, or the company's stockholders. In addition, the practical limits of media coverage may limit how many of these audiences you can address. Another complicating factor is that the interests of the audiences may not only be different but contradictory! Shareholders may need to know that the remediation will cost $50 million, and that between reserves and insurance it is covered. On the other hand, broadcasting coverage may only encourage litigation and holding out for higher settlements. Determining audience and message priorities requires a diverse range of corporate resources to ensure that every interested constituency's position is represented and considered.

B.   WHAT ARE THE KEY POINTS FOR EACH OF THE DIFFERENT PUBLICS?

These key points are the sound bites, the encapsulation of everything you want to say to each audience distilled to no more than 15-second messages. Based on our experience, it is best to have no more than three such points. In this era of information overload, an audience (regardless of its sophistication or concern) has an attention span for no more than three, hopefully related, ideas.

C.   WHEN DOES THE COMPANY FIRST COMMUNICATE?

When should you start communicating with your publics: after you are sued, after you sign a consent agreement, or after you are fined by a regulatory authority? Is it when pickets gather at the site or at the headquarters? Is it when you first decide that the site must be remediated? Is it when the first tests of the site come in suggesting that there is a toxic waste problem—actual or potential? Sooner is usually better than later, particularly if the company wants to be the first to establish the issues agenda and demonstrate issue control. Starting with control is easier than winning it later.

D. WHAT VEHICLE SHOULD BE USED TO COMMUNICATE?

Clearly, the answer to this question will differ for each circumstance. However, McLuhan was right: the medium is the message. A press conference usually means real trouble (or a final resolution). Interviews with experts usually mean you are reasonably confident of your position. Having a well-prepared spokesperson handling telephone inquiries usually communicates that the company is handling this "in the ordinary course of business." A press release on Friday afternoon at 5:00 p.m. says, "it is real bad news but we are too arrogant to discuss it."

E. WHO SHOULD DO THE COMMUNICATING?

The CEO clearly has the most currency, but this currency is used up very rapidly in a public controversy, particularly as new facts emerge that appear to contradict early statements. However, you surely will want the CEO up front if, in fact, the site has caused actual health problems. Only the CEO is acceptable in these circumstances for expressing the corporation's regret and its commitment to leading a resolution. The operating officer with direct responsibility for the particular issue or site being remediated is the highest-level executive who should speak for the company on most developments. However, in most circumstances, a well-trained public relations professional is the right person to be speaking to media, while a professional in community relations should be the one-to-one contact with concerned community organizations.

Each person with communications responsibilities must be trained, both for style (to ensure that they achieve the level of "performance" that television has trained the audience to expect of such people), and, just as important, to assure consistency of message in every media opportunity.

There is no excuse for using either outside legal counsel or outside public relations counsel as the company's spokespersons. Use of these messengers sends a bad message—that the company is wrong, incompetent, and needed to hire people to protect it. However, legal counsel is necessary for assisting in the development and/or review of key points and responses to inquiries—to ensure consideration of the company's legal position and the professionalism of message presentation. Nevertheless, their work must be invisible to the news media.

F. WHAT IS THE APPROPRIATE LEVEL OF BACKGROUND FOR EACH AUDIENCE?

The public generally does not need or want much, but just enough to know whether your key points are credible. An elected or regulatory official, on the other hand, may need a great deal of background, both to make required decisions, but, frankly, also to prove to his or her constituents "I am informed." Regardless of the level of background, the communication must be appropriate to the particular key point or development. Do not walk "into jail" by leading a constituency into areas for which you are not yet prepared.

G. FROM WHERE DO YOU COMMUNICATE?

From where do you communicate? Does the company send spokespersons to the site for interviews? Should the company invite media to its headquarters? Many environmentally impaired properties are unattractive, at best, and will clearly focus public attention on the question "what happened here?" The site also may be trafficked in a manner that suggests corporate disorder. Company headquarters on the other hand may look "plush," thereby providing opportunities for the media to use signage to remind the public of the name of the company under fire and, again, may be the scene of uncontrollable human traffic. It is the visual image and not anything that may be said that is critical.

H. ARE AUTHORITIES LIMITING COMMUNICATIONS (EITHER CONTENT, TIMING, SCOPE, OR MEDIA), AND SHOULD THE COMPANY COOPERATE WITH THESE LIMITATIONS?

A regulatory authority may be taking its time reviewing a test result while the public is accusing the company of "a cover-up" for not disclosing it. Does the company wait? The Department of Health may ask you not to describe symptoms of exposure, for fear of generating a "placebo effect" that will overload hospitals or distort data. Is it an exercise of corporate responsibility if early treatment can prevent the development of a more serious condition? The local police may want to order pickets away, because they are "a threat to traffic." During such action, the media will be asking whether the company will support this "outrage against the First Amendment."

I. WHAT IS THE DECISION OR APPROVAL PROCESS FOR STATEMENTS THAT THE COMPANY MAY MAKE?

If a remediation contractor has an accident on site that injures workers and may have an environmental impact, then who speaks—the contractor or the company? If the CEO is flying to a utility conference in New Zealand, who will approve the news release responding to a charge that 50 people developed cancer as a result of exposure to the site? Will you wait until the CEO is available? You have promised a communication to the neighbors, but the state environmental agency is holding up the review. Again, must you wait?

There are no easy answers to any of these questions, yet, at the end of the day, the most damning comment anyone can make is "if only we thought of." This is what decision priorities are really about, creating a process that ensures all issues, and all points-of-view on each issue, are considered by the decision-makers.

V. DEVELOPING A COMMUNICATIONS STRATEGY AND PLAN

Much of the foregoing, of course, is directed toward putting in place mechanisms to anticipate and respond to emerging situations from the time a site is identified as one with environmental problems through the various steps of a remediation. These mechanisms are important to be sure. Yet more important is developing a thorough communications strategy and plan coincident with the development of a management strategy for the impaired property itself.

A communications plan should begin with a very realistic situation analysis of how the site and the company's stewardship of it will be understood by the community (and particularly by such intermediaries to the community as its elected representatives and media) once the community is informed of the nature and scope of its contamination. While it can be assumed that intermediaries will be fair, such an analysis will be less than useful as a starting point for strategy development if it does not consider all possible critiques of the company and the site. The assumption of "fairness" does not necessarily reflect what will happen. Strategy development generally, and situation analysis specifically, must provide management with an opportunity to assess the effectiveness of its case against those cases that will be advanced by interest groups and others who want to use the site as a focal point for their agendas.

Situation analysis also must provide management with an opportunity to develop an understanding of how specific developments and specific issues are likely to evolve into public consciousness, and how these developments and issues will be played out in media coverage of the site and its remediation. We recommend that this be done through the development of anticipatable scenarios of the following types:

- "sudden" events, such as the identification of specific injuries or an accident during the remediation process that does have an environmental or health impact;

- "planned" events, such as an attack on the company's or regulator's stewardship by an activist group or counsel to a group of actual or prospective plaintiffs; and

- "issues" events, such as legislative hearings, in-depth stories by print or electronic media and so forth.

Scenario presentations must include several elements, including:

- a summary description of the event;

- an estimation of how the event will be brought to the attention of media;

- an estimation of how the story would be reported absent comment from the company or its supporters;

- an identification of persons or groups likely to be sources to the media and their "clout" or credibility;

- a summary presentation of what the company's going-in position and key points are likely to be;

- an identification of people or organizations outside of the company who might support the company's position; and

- preliminary, anticipated decisions using the Decision Priorities process.

The Going-in and Key Points sections are critical. First, they help the company understand its relationship to the site and the likely community response.

Second, when an anticipated scenario becomes a reality, common media practice is to come to the company for comment after it has talked to every other source, and to do so probably no more than an hour before "deadline." Having prepared going-in positions expedites development and approval of actual messages. It also expedites overall decision processing. Therefore, the company's chances for getting the right messages out are improved, and the confidence with which this is done will be reflected in the demonstration of responsible management action, which is at the heart of communicating with the media and the community at large.

Many management groups require that an anticipatory communications plan contain a detailed set of Questions and Answers (Q&A). While this form of planning can be useful, the reality is that every question has to be seen as an opportunity to present one of the company's Key Points about the site and its remediation. As Henry Kissinger said at a press conference, "I have the answers. Do you happen to have any questions?" In this construct, the Q&A section of the communications plan is best seen and used as a training device. Its purpose is to guide management and spokespersons how to use questions to bridge to the messages (Key Points) that the company knows it must send to its audiences given the realistic expectation of a less than 15-second sound bite.

These elements are at the heart of a strategy that is responsive. It is usually best to develop the responsive elements of the strategy first because it is more likely than not that public discussion of a site and its problems will begin before the remediation plan has even begun to evolve.

This is not to say that management should not have a proactive strategy as well, one that attempts to control the discussion of the site even before the first community or media interest is expressed. However, it is critical to remember that just as in a responsive program, proactive communication has to focus on the four communications priorities: Define the Risk, Describe the Actions That Mitigate the Risk, Identify the Cause of the Risk, and Demonstrate Responsible Management Action.

Taking the initiative both in defining the risk and identifying its cause is not without controversy as a management decision. The advantage here is entirely from the public relations/communications perspective. By taking the initiative and assuming the absolute honesty that such action requires, the corporation has:

- defined the key issues for all interested parties;

- presented itself as a company that willingly takes responsibility for its actions (or those of its predecessors) as well as for the community affected by them; and

- demonstrated the type of pro-activity expected of contemporary corporations, particularly as such corporations are defined for the public by such intermediaries as reporters and financial analysts.

The downside of taking the initiative is that some will argue it is tantamount to admission of liability for legal or insurance purposes, and may require identifying and "punishing" either a manager or a management system in identifying the cause of the site's pollution.

The reality is that liability is going to exist whether or not the company is the first to disclose the problem. In fact, one could argue that continuing not to discuss the site may exacerbate or prolong exposures with a strong likelihood of increasing liability. Whether or not the company adopts a proactive strategy (if in fact it is responsible for the site and for its remediation), at some point it will have to identify why the situation exists and the manager or process that was responsible for it. Whether this occasion happens now or much later is irrelevant to the ultimate success of the remediation, or to its ultimate cost. However, particularly in the era of omnipresent "we-never-close" media, the public relations cost of apparent delay may prove insurmountable.

The final consideration in developing a strategy—responsive as well as proactive—is for management to have agreed in advance, as to what constitutes a successful remediation project:

- Is it simply limiting the amount of public controversy a remediation usually engenders? That can be a goal, although usually not a very realistic one.

- Is it achieving a remediation that returns the property to some use or no use, but at the lowest possible financial cost? This end point can be both realistic and achievable, but will not provide the company with the paradigm values that may enhance its community, business, or even investment reputation (as Johnson & Johnson's very successful management of the Tylenol® tampering so clearly achieved for that brand and company).

- Alternatively, should the goal be remediating the site in a way that returns the site to a use that is valuable to the community, provides a measure of financial return to the company, and demonstrates leveragable management characteristics? Again, this objective can be achieved but, under certain circumstances, may not be realistic because of the complex of issues and political, social, and cultural interests that may have to be brought into some form of consensus.

Remediation that is oriented toward achieving a community value, for example, may spark controversy within the community itself as to which values need to be developed: business and economic, social and cultural, and even environmental and recreational. Similarly, remediation oriented towards creating a financial value may spark controversy over who in the community is going to receive those values, or whether it is even appropriate for a corporation to achieve a financial return given the environmental and possible health cost its site generated for the community as a whole.

It is rare for companies to exercise the will to tackle these incredibly complex issues, and to commit the range of managerial, engineering, legal, communication, and financial resources necessary to accomplish the valuable use of remediation successfully. Nevertheless, our experience is:

- if a company is committed to a strategy from the time the decision is first made to conduct a remedial investigation;

- if that strategy reflects the priorities of the public and the reasonable expectations of the media;

- if there is a process in place that is not only communicable in its own right but also evident in every act and every communications the company undertakes; and

- if there is a continued focus on both the reality and appearance of management controlling the action in a manner responsible to the interests of the company and the community, then, and only then, does such a remediation become realistic as well.

It is hard, now, to determine whether such a commitment and such an investment will pay off. To our knowledge, no such completed remediation has had enough time to be tested for whether all of the anticipated values can be achieved. However, there are examples from other industries and other circumstances that suggest that aggressive and proactive remediation strategies can enhance a panoply of positive values:

- The Tylenol® example is one (although, in that case the strategy was entirely responsive until J&J's management actually re-launched the brand).

- Owens-Illinois' proactive strategy to cleanup an abandoned dock area it had used on the Maumee River, creating Seagate Plaza and leading to the renaissance of Toledo's downtown, in turn producing economic value to the community and to O-I in the overall valuation of the company and its real estate holdings.

- The remediation of former MGP plants on the waterfronts of Milwaukee and Erie, Pennsylvania also are showing signs of generating these broad values, with leveragable credits going to the owners' financial as well as reputational accounts.

In every instance, these seemingly successful programs reflect the commitment of senior management. They also demonstrate thoroughly planned communications that directly involved all elements of the community in the remediation process. In so doing, attention was constantly focused on the fact that the site, the public discussion of issues related to the site, and the remediation itself were subject to an understandable management process. That process, which is the substance spoken of at the beginning of this paper, is a process that is always focused on addressing the priorities the public has been trained to expect.

Appendix D

RISK-BASED ANALYSIS WORKBOOK

Kurt A. Frantzen, Judy Vangalio, and Cris Williams

The Risk-Based Analysis Workbook is a tool to help managers plan and implement an environmental risk management project following the approach described in this book. This appendix has a two-fold objective: as a manual summarizing the approach (with a checklist and guiding framework for performing each of the five steps of Risk-Based Analysis as shown in Table D.1) and as a record of work related to a project.

Table D.1

Risk-Based Analysis:
Five Progressive, Knowledge-Building Value Points

Problem Formulation

First, define the problem; specify needed resources, deadlines, and scope. Use conceptual models to guide definition of source (cause), effect, and the many influencing factors. Establish the boundaries and operational context of the problem and the associated impairment or risk issue(s). Develop a preliminary model of the decision-making process and identify data needs to inform that process and define the necessary quality of data (*i.e.*, if you collect or calculate it, will it convince).

Situational Analysis

Identify, understand, and integrate the needs and objectives of others within the regulatory, political, and socioeconomic aspects of the property and their roles in risk management decision-making.

Risk Assessment

Quantify and qualify the nature, frequency, and intensity of risk. Set the scientific data and findings in redevelopment/reuse contexts.

Risk Management Option Development

Depending on the problem and its situational context, address what options are available to scientifically and justifiably explain away the reputed risk or impairment, cut-off exposure pathways (and therefore risk), or permanently reconstruct the "risk system" (*i.e.*, source or effect) so that it no longer exists, or is quantitatively reduced in magnitude by a significant and sufficient degree. In addition, to help influence outcome options, develop your risk mitigation scenarios within the context of redevelopment and economic revitalization. It may even be worthwhile to develop a short- and long-term amortization of risk over a sufficiently long planning horizon to better contain costs and land use.

Risk "Argument"

In this step, develop a convincing communications approach to achieve optimal, "mutual gain" solutions by integrating property value, environmental risk or impairment, the situational context, risk-management options, and decision-making frameworks. The risk information and preferred risk management option are formulated within a communications program by which it is presented and, ultimately, negotiated into an approach acceptable to all.

Table D.2
Problem Formulation
Background Information Checklist

<u>Property Information</u>

Property Name:
Property Address:
☐ Tax ID

☐ Chain of Title (50-year coverage with all instruments)

☐ Zoning Class

☐ Property Locator Maps
☐ Tax Assessor's Map
☐ Property Map(s) including:
    ☐ Legal Description
    ☐ Meets and Bounds
    ☐ Easements (on and around) including
        ☐ Sanitary   ☐ Storm Drainage
        ☐ Water   ☐ Cable
        ☐ Electric   ☐ Ingress/Egress
        ☐ Gas   ☐ Telephone
        ☐ Other
    ☐ Site Plan

<u>Current Area Maps / Data</u>

☐ Topographic   ☐ Transportation
☐ Drainage   ☐ Geological
☐ Soils Classification   ☐ Hydrogeological
☐ Flood   ☐ Surface water / watershed
☐ Wetlands
☐ Current Covertype
☐ Demographics
☐ Land Use (site and surrounding area; current and future/proposed changes, development
    plans)

<u>Historical Maps</u>

☐ Sanborn Insurance Maps
☐ Topographical
☐ Other:

<u>Climatic Data</u>

☐ Annual Temperature Data
☐ Precipitation Data
☐ Wind Rose
☐ Other:

Table D.2 continued

Photography

☐ Aerial Photography
    ☐ Historical   ☐ Current
☐ Surface Photography of Property and Structures
    ☐ Historical   ☐ Current

Facility Information

☐ Building/Facility Layout Sketch
☐ Industrial / Manufacturing Processes (type and description)
☐ Permits
    List all permits:

Site & Building Reconnaissance

☐ Site History
☐ Current Occupancy Status
☐ Site Walkover Report   ☐ Property Condition Assessment
☐ Site Access and Transportation Infrastructure
☐ Disposal of Waste Generated On-Site
☐ Storage Tanks and Pipelines   ☐ Utilities   ☐ Fuel / Heating
☐ Other:

Municipal Review

☐ Assessor's Office
☐ Ownership History
☐ Fire Department
☐ Building Department
☐ Board of Health
☐ Public Works Department
☐ Library

Regulatory Review

☐ USEPA
☐ State
☐ Local
☐ Other:

Receptors / Resources

☐ Nearest Residence
☐ Sensitive Receptor Identification and Location
☐ Community Health Status Reports
☐ Natural Heritage Program   ☐ Biotic/Vegetative Resources ☐ R/E/T Species
☐ Coastal Zone / Significant Habitat Information
☐ Well Survey

Table D.3
Problem Formulation
Initial Problem Statement

Presenting Problem

Briefly state problem in general terms and known facts as to what is/is not, or should/should not be happening. Then describe the problem graphically using a Conceptual Site Model (see Table D.4) and diagnose the problem by developing a Conceptual Risk System Model (see Figure D.1).

Bound the Problem	Who	What	Where	When	Comments
Describe Distinctive Characteristics of the Problem					
Scope of Each Characteristic					
Magnitude of Each Characteristic					

Table D.4
Problem Formulation
Conceptual Site Model

Historical

This section of the model should provide maps, (aerial) photographs, sketches, or other such information to provide an understanding of the property under former land use. The purpose is to provide a spatial understanding of the location of historical discharges and other sources of hazards in relation to relevant landmarks, landforms, or other items relevant to the property and abutting property.

Current Land Use

This section of the model provides a base map of the property as it is currently configured and zoned, with structures. It should present topographic relief, show the directions of drainage, surface water flow, and groundwater flow (regional and/or specific, as known). This portion of the model should include a Wind Rose indicating the annualized prevailing wind speed and direction.

Future Land Use

This portion of the model should diagram expected land use changes for the property and abutting properties over a relevant time horizon to inform the process.

Environmental Features

This portion of the model provides biological, ecological, and critical environmental information about and around (in a 0.5- to 2-mile radius) the property. It should identify flood and coastal zones, historical sites and areas, critical habitat, rare/endangered/threatened species locations (if known), and the approximate location of on- or off-site human receptors, especially sensitive ones (e.g., children, elderly, informed).

Locational Maps

Using simple maps indicate the general local and state location of the property.

Purpose is to describe the problem graphically in terms of space, time, and relationships of important abiotic, biotic, and cultural items and factors present at or in proximity to the impaired property of interest.

Media	Off-Site	On-site	Off-Site
Air (ambient) (indoor)			
Soil/Sediment (surface <3 in) (intermediate 3 in - 15 ft) (subsurface >15 ft)			
Water (ground or surface)			

Media										
Exposure Route										
Pathway #										
People										
Current — On-Site										
Current — Off-Site										
Future — On-Site										
Future — Off-Site										
Biotic Receptors										
Plants — On-Site										
Plants — Off-Site										
Animals — On-Site										
Animals — Off-Site										

Figure D.1 Problem Formulation, Step One: Conceptual Risk System Model. This model describes how things of concern are thought to be moving through the environment resulting in opportunities for exposure for people, places, resources, and/or things.

Table D.5

Problem Formulation
Defining the Decision: Issues Analysis

Priority	Issue	Type	Scope

Issue Types:
In this step, identify those issues from the Conceptual Models that are objective, tangible, and most likely to be qualifiable and quantifiable. Issue types and specifics are shown in Table D.6.

Table D.6

Problem Formulation
Defining the Decision: Objective Issue Typology

Media of Concern

☐ Air
    ☐ Ambient   ☐ Indoor
☐ Soil
    ☐ Surface (0 –   inches)   ☐ Sub-surface (depth =   feet)
☐ Water
    ☐ Groundwater   ☐ Surface water
☐ Sediment
    ☐ Surface (0 –   inches)   ☐ Sub-surface (depth =   feet)
☐ Biota (identify below)
    ☐ Flora   ☐ Fruits   ☐ Vegetables (root)
    ☐ Bird   ☐ Wildlife   ☐ Vegetables (leaf)
    ☐ Fish   ☐ Benthos   ☐ Soil Organisms

Chemical Constituents

Class   **Specific**
☐ VOCs
☐ Base / Neutral Extractable
☐ Acid Extractable
☐ Metals
☐ Inorganics
☐ Pesticides / PCBs
☐ Mixtures

Physical Stressors

List and describe:

Biological Agents

List and describe:

Ecosystem Type

Describe the type of ecosystem at the site including any sensitive habitat or other special features:

Table D.6 continued

Possible Human Receptors

Type	On-site	Off-site	Comments
☐ Residential	☐	☐	____
☐ -Adult	☐ -Child	☐ -Other	____
☐ Gardener	☐	☐	____
☐ Subsistence Fisher	☐	☐	____
☐ Recreational	☐	☐	____
☐ -Adult	☐ -Child	☐ -Other	____
☐ Trespasser	☐	☐	____
☐ -Adult	☐ -Child	☐ -Other	____
☐ Commercial Worker	☐	☐	____
☐ -Indoor	☐ -Outdoor	☐ -Other	____
☐ Industrial Worker	☐	☐	____
	☐ -Maintenance		____
☐ Construction Worker	☐	☐	____

Additional description of receptors, as necessary: ____

Regulatory Issues

☐ Local

    ☐ Permitting   ☐ Zoning   ☐ Other (____)

☐ State

    ☐ Air   ☐ Water ☐ Hazardous Waste ☐ Other (____)

☐ Federal

    Describe (____)

Organizational Issues

☐ Property value
    Describe (____)

☐ Operational value of property
    Describe (____)

☐ Transactional value
    Describe (____)

Table D.7
Problem Formulation
Defining the Decision

Restate Formulated Problem

Briefly, state problem in terms that are more specific, following what is graphically presented in the concept models (Table D.4 and Figure D.1). Next, formally state what needs to be determined and why, in non-binary terms. Recognize multiple decisions and stages within the remedial and/or redevelopment process. Prioritize and describe logical relationships between any of the decisions. Additionally, identify what you are trying to achieve, preserve, and avoid as tangential problems by whatever is decided.

Decision(s)	Stage	Priority	Achieve	Preserve	Avoid	Comment

Table D.8

Problem Formulation
Informing the Decision

Data Quality Objectives Process Summary

☐   **State the Problem**

This is the purpose of the program (Step 1 of Problem Formulation, see Tables D.1 and D.7 and Figures 4.2, 4.3, and D.1).

☐   **Identify the Decision(s)**

What is the actual decision (which may require several interrelated decisions)? (Step 2 of Problem Formulation; see Tables D.5, D.6, and D.7)

☐   **Identify Inputs to the Decision(s)**

What information is required to make the decision(s)? What information is already in-hand; what are the data gaps? What are the information collection steps, sources, and procedures? (Step 4 of Problem Formulation)

☐   **Define the Boundaries of the Decision(s)**

It is necessary to specify the boundary limits of any information activity in terms of the property and the nature of the environmental impairment under consideration. These limits include property boundaries, definition of different on-site areas (for decision-making purposes), off-site activity locations and limits, and temporal issues (operational activities, seasonal or tidal timing, etc.). In essence, one is setting the scale of the decision in terms of spatial area, volume, and/or time frame. Furthermore, it is appropriate and necessary to define the cost and schedule constraints as this affects decision inputs and decision error limits. (Step 1 of Problem Formulation, see Tables D.1 and D.7 and Figures 4.2, 4.3, and D.1)

☐   **Develop Decision Rule(s)**

This is simply the development of a statement specifying those conditions that would cause a decision between several outcomes or courses of action. (Step 2 of Problem Formulation; see Tables D.5, D.6, and D.7)

☐   **Specify Limits on Decision Errors**

This statistical specification can be qualitative or quantitative and indicates the level of accuracy and precision necessary for the data selected for the decision. This is a necessary step because of the natural variability of environmental conditions, sampling, and analysis; errors in sampling and analysis; and uncertainty. (Step 2 of Problem Formulation; see Tables D.5, D.6, and D.7)

☐   **Optimize Design of the Data Collection and Analysis Activity**

Considering the definition of the problem, the nature of the decision, and the limits of the effort (including decision errors) it is possible to develop a resource- and cost-effective approach for obtaining the information. (Step 4 of Problem Formulation)

Source: USEPA, 2000

Table D.9

Problem Formulation
Planning the Rest of the Process

☐ Identify the members of your own internal knowledge-based network of other corporate functions and engage them in the process.

Legal   _____
Environmental   _____
Corporate Relations   _____
Financial   _____
Facilities   _____
Other   _____

☐ Educate and mobilize them into the process so as to integrate technical, legal, and regulatory relations issues to decide how to operate the remedial/corrective action process and under which appropriate regulatory domain (local, state, or federal), that is if you have the opportunity to choose.

☐ Develop an appreciation of the impaired property as an asset and the community and marketplace in which it is located. This is essential to applying an integrated approach to the economic redevelopment of environmentally impaired properties. All other issues pertaining to a site—whether it be surrounding infrastructure, environmental conditions, demographic profiles, etc.—are weighed in terms of how they impact the economic candidacy of a potential Brownfields redevelopment project. The strength of candidacy centers on returns-on-investment associated with achieving regulatory compliance while designing risk through the site remediation process and predicated on reuse concepts of the property. While this logic may be readily apparent, values placed on properties frequently do not account for variables that have real limitations associated with restoration costs. Even in "upside-down" cases, economic returns can be enormous—not within the site unto itself, perhaps, but in relationship to its impact on surrounding sites and/or overall site vicinity (Ackerman *et al.*, 1998 and Ackerman and Soler, 2000).

☐ Engage the finance and risk management groups to help determine opportunity costs and liabilities, as well as coverage and reimbursement options as suggested in Figure 3.7.

☐ Plan for the integration of external technical, legal, and other resources for completing necessary components of the many and varied follow-on processes.

☐ Establish the planning horizon and develop a timeline to identify crucial deadlines and critical sequencing issues (Bélanger and Craig, 1999).

☐ Develop a specific action plan.

Table D.10

Situation Analysis
Defining the Decision: Players

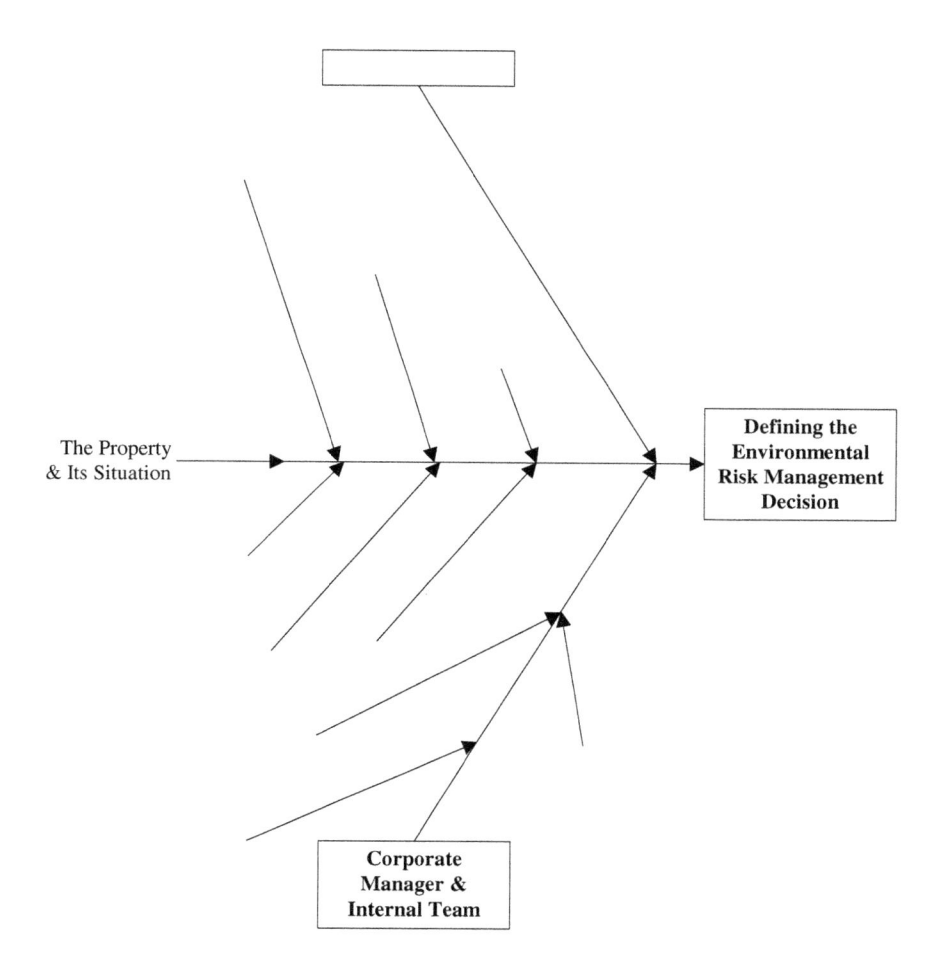

Name/list those involved or likely to be involved in resolving the problem and identify their roles. Identify primary and secondary decision-makers and decision-influencers by proximity to the decision(s). Identify other possible stakeholders and constituents as they arise during the process.

Table D.11
Situation Analysis
Defining the Decision: Issues

Priority	Issue	Type	Scope

Issue Types:
In this step, identify those issues from the Conceptual Models that are objective, tangible, and most likely to be qualifiable and quantifiable. (Issue types and specifics are shown in Table D.12.)

Table D.12

Situation Analysis
Defining the Decision: Subjective Issue Typology

<u>**Political Issues**</u>

☐ Community
    Describe (_____)
    Importance (_____)   Significance (   )

☐ Local
    Describe (_____)
    Importance (_____)   Significance (   )

☐ State
    Describe (_____)
    Importance (_____)   Significance (   )

<u>**Image Issues**</u>

☐ Internal
    Importance (_____)   Significance (   )

☐ External
    ☐ Political   ☐ Regulatory   ☐ Community
    Importance (_____)   Significance (   )

<u>**Value Issues**</u>

☐ Internal
    ☐ Financial   ☐ Operational   ☐ Transactional
    Describe (_____)
    Importance (_____)   Significance (   )

☐ External
    ☐ Economic
    Describe (_____)
    Importance (_____)   Significance (   )

    ☐ Other
    Describe (_____)
    Importance (_____)   Significance (   )

    ☐ Quality of Life
        ☐ Aesthetic   Describe (_____)
           Importance (_____)   Significance (   )
        ☐ Community Living Conditions   Describe (_____)
           Importance (_____)   Significance (   )
        ☐ Cultural / Historical Heritage   Describe (_____)
           Importance (_____)   Significance (   )
        ☐ Other   Describe (_____)
           Importance (_____)   Significance (   )

<u>**Uncertainties**</u>

☐ Other
Describe (_____)
Importance (_____)   Significance (   )

Table D.13

Situation Analysis
Problem Formulation Update

Update the Presenting Problem and Related Problems

Briefly, re-state the problem in terms that are more specific based on an understanding of the players and issues. Reformulate as necessary the concept models (Table D.4 and Figure D.1). When considering the decision(s), ask yourself the following questions. What are the strengths and weaknesses of your position? Who are your allies and who is against you, and what are their various strengths and weaknesses? How can you create partnerships and build a mutual gain that can yield lower cost and liability for your firm, a safe and improved environment and community for stakeholders, another positively resolved impaired site for the regulators, and reduced risk for all? What are the threats of possible alternatives?

Decision(s)	Stage	Priority	Achieve	Preserve	Avoid	Comment

☐ Update management strategy
☐ Communications plan

Table D.14
Managing the Risk Assessment
Rules-of-Thumb for a Baseline Assessment

Conceptual Site Model: Evaluate available data and develop a well-defined CSM as early as possible. The CSM is a three-dimensional "picture" of site conditions that illustrates contaminant sources, release mechanisms, exposure pathways, migration routes, and potential human and ecological receptors. The CSM documents current and potential future site conditions and is supported by maps, cross sections, and site diagrams that illustrate what is known about human and environmental exposure through contaminant release and migration to potential receptors. The CSM is initially developed during the scoping phase of the RI/FS and should be modified as additional information becomes available.

Exposure Pathways: Evaluate all relevant exposure pathways related to the site (*e.g.*, direct ingestion, inhalation), for both current and reasonably anticipated future land uses as well as current and potential future groundwater and surface water uses.

Data Needs: Collect sufficient chemical concentration data from each relevant medium to adequately characterize the nature and extent of contamination and to develop sound estimates of risk associated with each exposure pathway.

Site-Specific Risk Calculation:
- Calculate the cumulative risks to an individual for chronic exposures, using reasonable maximum exposure (RME) assumptions by combining a statistically sound, arithmetic average, exposure-point concentration with reasonably conservative values for intake and duration.
- Use the most current toxicity values provided by the appropriate regulatory agency, the Integrated Risk Information System (IRIS, USEPA, 2001b) or the Health Effects Assessment Summary Tables (HEAST, USEPA, 1997g).
- Include estimates of risk for current and reasonably anticipated future land uses and potential future groundwater and surface water uses, without institutional controls. The baseline risk assessment is essentially an evaluation of the "no action" alternative (*i.e.*, an assessment of the risk associated with a site in the absence of any remedial action or control). While institutional controls do not actively clean up the contamination at a site, they can control exposure and, therefore, are considered limited action alternatives that may be evaluated during the remedy selection process.
- Include a discussion that identifies major sources of uncertainty or variability and their influence on the risk estimates. Probabilistic methods may aid in evaluating uncertainty at some sites.

Other Measures of Risk: Other measures of risk (*e.g.*, central tendency) can be used to describe site risks more fully. However, RME risk generally should be the principal basis for evaluating potential risks at Superfund sites.

Exposed Populations: The risk analysis should clearly identify the population, or population sub-group (*e.g.*, highly exposed or susceptible individuals), for which risks are being evaluated.

Ecological Risk Assessment: Include an assessment of ecological risk in the baseline risk assessment in order to support EPA's mission to protect the environment. A screening ecological risk assessment generally should be conducted to identify those chemicals, media, and portions of the site requiring more detailed study and analysis. Use site-specific toxicity tests, field studies, and food-chain models whenever appropriate.

After USEPA, 1997c

Table D.15
Managing the Risk Assessment

<u>**Risk Assessment Team**</u>

Corporate Management

Environmental Staff

Legal

Relations

Technical Consultants

Other Support

<u>**Risk Characterization**</u>

General Types	Specific Types	Qualitative Description	Quantitative Description	Rank Importance	What is Significance Level?	Uncertainty
Public Health						
Ecological						
Socio-Cultural						
Socio-Economic						

Table D.16
Risk Management Options
Rules-of-Thumb

Basis for Action: Response action is usually warranted if at least one of the following conditions is met. If Cumulative Excess Carcinogenic Risk to an individual is greater than 10^{-4} (applying Reasonable Maximum Exposure [RME] assumptions for current or reasonably anticipated future land use). If Non-carcinogenic Hazard Index is greater than one (applying Reasonable Maximum Exposure assumptions for current or reasonably anticipated future land use). If the contaminants may cause adverse impacts to the environment. If chemical-specific standards, or other measures that define acceptable risk levels, are exceeded and exposure to contaminants above these acceptable levels is predicted for the RME (*e.g.*, drinking water standards exceeded in groundwater when it is a current or potential source of drinking water; or water quality standards are exceeded in surface or groundwater supporting the designated uses of those waters [*e.g.*, supporting aquatic life]).

Preliminary Remediation Goals (PRGs) for Carcinogens: In the absence of ARARs for chemicals that pose carcinogenic risks, PRGs generally should be established at concentrations that achieve 10^{-6} excess cancer risk, modifying as appropriate based on exposure, uncertainty, and technical feasibility factors.

PRGs for Non-Carcinogens: In the absence of ARARs for chemicals that pose non-carcinogenic risks, generally establish PRGs at concentrations where the Hazard Quotient is one. Cumulative non-cancer risks are determined by adding Hazard Quotients for chemicals with the same toxic endpoint or mechanism of action. In establishing PRGs for chemicals that affect the same target organ/system, divide the PRGs for individual chemicals by the number of chemicals present in this group.

Chemical-Specific ARARs: If a single ARAR for a specific chemical (or in some cases a group of chemicals) is used to define the acceptable level of exposure, then compliance with the ARAR is generally protective even if it is outside the risk range (unless there are extenuating circumstances, such as exposure to multiple contaminants or pathways).

Background Concentrations: USEPA does not generally clean up below natural background levels. However, where human-made background levels exceed acceptable risk-based levels, and EPA has determined that a response action is appropriate, EPA's goal is to develop a comprehensive response to address area-wide contamination. This will help avoid the creation of "clean islands."

Selecting Remedial Action: In the absence of ARARs, remedies should reduce risks from carcinogenic contaminants such that the Excess Cumulative Individual Lifetime Cancer Risk for site-related exposures falls between 10^{-4} and 10^{-6}. EPA prefers cleanups achieving the more protective end of the risk range (*i.e.*, 10^{-6}). (NOTE: The upper boundary of the risk range is not a discrete line at 10^{-4}, although EPA generally uses 10^{-4} in making risk management decisions. A specific risk estimate around 10^{-4} can be acceptable, if justified based on site-specific conditions.) For non-carcinogens, remedies generally should reduce contaminant concentrations such that exposed populations or sensitive sub-populations will not experience adverse effects during all or part of a lifetime, incorporating an adequate margin of safety (*i.e.*, Hazard Index = 1.0).

Timing: A "phased approach" to site investigation and cleanup generally will accelerate risk reduction and provide additional technical site information on which to base long-term risk management decisions. Use phased cleanup approaches wherever practicable (40 CFR 300.430(a)(1)(ii)(A)).

After USEPA, 1997c.

Table D.17

Integrating Remedial/Corrective Action and Redevelopment Plans
(Following the Brownfields revitalization process)

To accomplish this integration, the following components at a minimum are fused throughout each phase along the critical path. Several items hearken back to previous Risk-Based Analysis steps; those critical to integration are discussed in more detail below.

☐ **Stakeholder Identification** – Ascertain all parties with a stake in the project.
☐ **Surrounding Infrastructure** – Evaluate the attractiveness and costs associated with maintenance/upgrades.
☐ **Public/Private Vested Interests** – Establish the public/private partnerships required to subsidize the site.
☐ **Preliminary Planning** – Incorporate environmental assessment results into preliminary redevelopment plans.
☐ **Site Investigations** – Environmental investigations are an aspect of redevelopment investigations.

☐ **Feasibility Studies** – Potential redevelopment plans are further integrated with site remedies in this phase of the process to create optimized, market-driven redevelopment plans. With the concurrence with the site owner or other appropriately designated party, the project team will need to present, review, and discuss each potential redevelopment plan with the stakeholders and selected consultants and/or contractors to enable those entities to evaluate each plan for compatibility with remedial action alternatives for the site.
☐ **Layout Plans** –Based on the results of previous activities modify and refine any or all of the potential redevelopment plans, as appropriate, to achieve more specified compatibility of the conceptual layout alternatives with remedial action options. These plans are conceptual master plans with key attributes such as: an overall reuse and/or subdivision schematic; a rendering of vehicular and pedestrian networks; plans of selected parcels showing proposed new building locations, roads, open space, and amenities; and a phasing plan, defining sequential action for specific segments of the property in conjunction with the remediation plan and schedule.
☐ **Generate Market Interest** – Depict environmental cleanup as a catalyst to economic redevelopment.
☐ **Opportunities and Constraints** – Optimize redevelopment opportunities with environmental strategies.
☐ **Risk Management** – Define risk in terms of remediation and redevelopment activities tied to future site use.
☐ **Cost Benefit Analysis** – Apply redevelopment expertise to assess risk and cost benefits of each strategy. This activity calls for evaluating the costs of remediation and site redevelopment. Appropriate consultants will need to provide qualified land development planning and engineering advice to stakeholders as an assessment of risks/cost-benefits of each alternative redevelopment scenario is conducted. The evaluation will be based on factors such as alternative alignments and layouts for "clean" utility corridors, building locations, and building foundations. Because of these assessments, any or all of the potential redevelopment plans may need to be modified and refined.

☐ **Communications** – Risk management should involve a communications strategy to maintain trust and credibility while building solid working relationships among all stakeholders.
☐ **Marketing** – Secure end uses and users to make environmental and economic rewards a reality.
☐ **Scheduling** – Implement task-specific timeframes with each project milestone. Time is money.

After Ackerman et al., 1998; used with permission.

Table D.18

Risk Arguments

The Rules

1. There is no winning or losing.
2. Don't argue with the lovers or haters.
3. If the media says it's a crisis, then it is!

Risk Communication Priorities

A. Define the risk for each audience.
B. Describe actions that mitigate the risk.
C. Identify the cause of the risk.
D. Demonstrate responsible management action.

Decision Priorities

1. To whom must we communicate and in what priority?
2. What are the key points for each different public?
3. When does the company first communicate?
4. What vehicle should be used to communicate?
5. Who should communicate?
6. What is the appropriate level of background for each audience?
7. From where do you communicate?
8. Are authorities limiting communications (either content, timing, scope, or media), and should the company cooperate with these limitations?
9. What is the decision or approval process for statements that the company may make?

☐ Management Plan

☐ Communications Strategy

Adapted from Appendix C by Ostrow.

Appendix E

ACRONYMS AND GLOSSARY

I. ACRONYMS

ADR	Alternate Dispute Resolution
AOPC	Agent of Potential Concern
ARAR	Applicable or Relevant and Appropriate Requirement
ASTM	American Society for Testing and Materials
ATSDR	Agency for Toxic Substances and Disease Registry
BNA	Bureau of National Affairs
BRA	Baseline Risk Assessment
BW	Body Weight
CARACAS	Concerted Action on Risk Assessment for Contaminated Sites
CBEP	Community Based Environmental Protection
CEO	Chief Executive Officer
CERCLA	Comprehensive Environmental Response, Compensation, and Liability Act
CERSA	Concise Environmental and Redevelopment Assessment
CFR	Code of Federal Regulation
CPA	Certified Public Accountant
CPSC	Consumer Product Safety Commission
CRSM	Conceptual Risk System Model
CSF	Cancer Slope Factor
CSM	Conceptual Site Model
CTE	Central Tendency Exposure
2-D MCA	Two-Dimensional Monte Carlo Analysis
DAD	Decide Announce Defend
DAI	Define-Agree-Implement
DES	Diethylstilbestrol
DoD	Department of Defense
DQO	Data Quality Objectives
ED	Exposure Duration
EE_{diet}	Estimated Exposure from Diet
EE_{soil}	Estimated Exposure from Soil

EE_{total}	Total Estimated Exposure
EE_{water}	Estimated Exposure from Water
ELI	Environmental Law Institute
EPA	Environmental Protection Agency
FASB	Financial Accounting Standards Board
FDA	Food and Drug Administration
FFRRO	Federal Facilities Restoration and Reuse Office
FS	Feasibility Study
HEAST	Health Effects Assessment Tables
HQ	Hazard Quotient
HRA	Health Risk Appraisal
IARC	International Agency for Research in Cancer
IR	Ingestion Rate
IRIS	Integrated Risk Information System
IRLG	Interagency Regulatory Liaison Group
IRM	Interim Remedial Measure
LMS	Linearized Multistage
LOAEL	Lowest Observed Adverse Effect Level
MCA	Monte Carlo Analysis
MCP	Massachusetts Contingency Plan
MEE	Micro-exposure Event Analysis
MGP	Manufactured Gas Plant
NAS	National Academy of Science
NCP	National Contingency Plan
NEPA	National Environmental Protection Act
NGO	Non-Government Organization
NOAEL	No Observed Adverse Effect Level
NRC	National Research Council
NTEC	Not to Exceed Concentration
O and M	Operation and Maintenance
OSHA	Occupational Safety and Health Administration
OSTP	Office of Science and Technology Policy
OSWER	Office of Solid Waste and Emergency Response
PAH	Polycyclic Aromatic Hydrocarbon
PCB	Polychlorinated biphenyl

PCCRARM	Presidential / Congressional Commission on Risk Assessment and Risk Management
PDF	Probability Density Function
PMF	Probability Mass Function
PRA	Probabilistic Risk Assessment
PRG	Preliminary Remediation Goal
Q & A	Questions and Answers
RAGs	Risk Assessment Guidance
RAIS	Risk Assessment Information Service
RBA	Risk-Based Analysis
RBCA	Risk-Based Corrective Action
RBSL	Risk-Based Screening Level
RCRA	Resource Conservation and Recovery Act
REC	Recognized Environmental Condition
RfD	Reference Dose
RI	Remedial Investigation
RME	Reasonable Maximum Exposure
SDA	Situation Definition and Analysis
SMDP	Scientific/Management Decision Point
SPHEM	Superfund Public Health Evaluation Manual
SSTL	Site-Specific Target Level
SUF	Site Use Factor
TDI	Tolerable Daily Intake
TQM	Total Quality Management
TRD	Toxicity Reference Dose
TRV	Toxicity Reference Value
USDA	United States Department of Agriculture
USEPA	United States Environmental Protection Agency
UST	Underground Storage Tank

II. GLOSSARY

Abiotic: The non-living component of an ecosystem.

Actual Risk: The potential for an adverse outcome.

Acute Exposure: A single exposure to a toxic substance, which results in severe biological harm or death. Acute exposures are usually characterized as lasting no longer than a day or a short time period relative to the life of the organisms experiencing exposure.

Acute Toxicity: The deleterious, often poisonous effect of a substance characterized as evoking biological harm, including death from a single exposure.

Agent of Potential Concern: A substance (chemical, physical, or biological) detected at a site that has the potential to affect human ands/or ecological receptors adversely due to its concentration, distribution, and mode of toxicity.

Assessment End Point: An explicit expression of the environmental value that is to be protected. The Assessment End Point is the product of the Problem Formulation phase of a risk assessment; it defines the focus of the investigation.

Bioavailability: The degree to which a material in environmental media can be assimilated by an organism.

Biotic: The natural or living component of an ecosystem.

Brownfields: Land requiring regeneration, land previously used for industrial purposes having various levels of site contamination and/or structures requiring demolition and decontamination activities.

Cancer Slope Factor (CSF): An estimate of carcinogenic potency determined using the linearized multistage model and high-dose animal (or human) carcinogenicity studies.

Carcinogen: Any agent for which the EPA has determined that there is sufficient evidence that exposure may result in continuing, uncontrolled cell division (cancer) in humans and/or animals.

Chronic Toxicity: The deleterious, often poisonous effect of a substance characterized as evoking biological harm, including death, from an extended exposure.

Cleanup: Actions taken to deal with a release or threat of release of a hazardous substance that could affect humans and/or the environment. The term "clean-up" is sometimes used interchangeably with the terms remedial action, removal action, response action, or corrective action.

Concentration: The relative amount of a substance in an environmental medium, expressed by relative mass (e.g., mg/kg), volume (ml/liter), or number of units (parts per million).

Conceptual Model: An illustration depicting relationships among human, ecological resources, and their physical/chemical environment. The conceptual model incorporates food web relationships, fate, and transport of chemical, and possible exposure routes.

De minimus Risk: Risks that are considered trivial under the law.

Dose: A measure of exposure. Examples include (1) the amount of chemical ingested, (2) the amount of chemical absorbed, and (3) the product of ambient exposure concentration and the duration of exposure.

Ecological Impact: The effect that a man-made or natural activity has on living organisms and their non-living (abiotic) environment.

Ecology: The relationship of living things to one another and their environment, or the study of such relationships.

Endangerment Assessment: A study conducted to determine the nature and extent of contamination at a site on the National Priorities List and the risks posed to public health or the environment. The EPA or the state conducts the study when a legal action is to be taken to direct potentially responsible parties to clean up a site or pay for the clean up.

Exposure: Co-occurrence of or contact between a stressor and a receptor.

Exposure Pathway: The course a chemical or physical agent takes from a source to an exposed organism. Each exposure pathway includes a source or release from a source, an exposure point, and an exposure route. If the exposure point differs from the source, transport/exposure media (*i.e.*, air and water) also are included.

Exposure Point: A location of potential contact between an organism and a chemical or physical agent.

Exposure Route: The way a chemical or physical agent comes in contact with an organism (i.e., by ingestion, inhalation, or dermal contact).

Exposure Scenario: A set of assumptions concerning how an exposure takes place, including assumptions about the exposure setting, stressor characteristics, and activities of an organism that can lead to exposure.

Fate and Transport: The destiny and movement opportunities of a contaminating substance. Chemical, biological, and/or physical processes may alter or degrade the parent material; water, wind, or biological agents may influence movement from one locus to another in some time interval.

Feasibility Study: Analysis of the practicability of a proposal. The feasibility study usually recommends selection of a cost-effective alternative. It usually starts as soon as the remedial investigation is underway.

Hazard: The likelihood that a substance will cause an injury or adverse effect under specified conditions.

Hazard Index: The sum of more than one hazard quotient for multiple substances and/or multiple exposure pathways. The HI is calculated separately for chronic, subchronic, and shorter-duration exposures.

Hazard Quotient: The ratio of an exposure level to a substance to a toxicity value selected for the risk assessment for that substance.

Hazardous Substance: Any material that poses a threat to human health and/or the environment.

Impact: A change in a condition of state.

Ingestion Rate: The rate at which an organism consumes food, water, or other materials (e.g., soil, sediment). Ingestion rate usually is expressed in terms of unit of mass or volume per unit of time (e.g., kg/day, L/day).

Lowest Observable Adverse Effect Level (LOAEL): The lowest level of a stressor evaluated in a toxicity test or biological field survey that has a statistically significant adverse effect on the exposed organisms compared with unexposed organisms in a controlled condition or reference site.

Measurement End Point: A measurable ecological characteristic that is related to the valued characteristic chosen as the assessment end. Measurement end points often are expressed as the statistical or arithmetic summaries of the observations that make up the measurement.

Media: Specific environmental compartments – air, water, and soil – that are the subject of regulatory concern and activities.

No Observed Adverse Effect Level (NOAEL): The highest level of a stressor evaluated in a toxicity test or biological field survey that causes no statistically significant difference in effect compared with the controls or a reference site.

Noncarcinogen: Any agent for which the carcinogenic evidence is negative or insufficient.

Parameter: Constants applied to a model that are obtained by theoretical calculation or measurements taken at another time and/or place, and are assumed to be appropriate for the place and time being studied.

Perceived Risk: How people are likely to view the circumstances associated with a property.

Physical Agents: Mechanical equipment, or machinery that has the potential to cause a disturbance or an adverse effect.

Probabilistic Risk Assessment: An estimate of adverse effects that incorporate statistical distributions for hazards and exposure estimates.

Problem Formulation: Defines the problem; specify needed resources, deadlines, and scope. Use conceptual models to guide definition of source (cause), effect, and the many influencing factors. Establish the boundaries and operational context of the problem and the associated impairment or risk issue(s). Develop a preliminary model of the decision-making process and identify data needs to inform that process and define the necessary quality of data (*i.e.*, if you collect or calculate it, will it convince?).

Receptor:
A human population, plant, animal, community of organisms, or ecosystem that is exposed to stressors in the environment.

Reference Dose (RfD): A provisional estimate (with about an order of magnitude uncertainty) of a daily exposure to a human population, including sensitive subgroups, that is likely to be without an appreciable risk of deleterious effects during a lifetime.

Remediation:
The act, processes, or activities associated with providing remedies for problems; typically in the context of contaminated site clean-up.

Risk:
The probability of that a substance (chemical, physical, or biological) will produce harm under specified conditions.

Risk Analysis:
A two-step process of evaluating (quantifying) risk(s) and making (policy or reuse) decisions based on the evaluation together with other input.

Risk Assessment: The process of estimating the likelihood that a given effect will result from a specific, presence, action, or activity. Where likelihood is a probability and interpreted as the portion or fraction of time a consequence might be observed. Concerning toxic substances, risk assessment involves determining the likelihood of release (exposure) and the resulting consequence (hazard).

Risk Assessor:
An individual or team with the appropriate training and range of expertise necessary to conduct a risk assessment.

Risk Characterization: The last stage of the risk assessment that describes the relationship of hazard and exposure as an estimate of risk.

Risk Communication: The exchange of information about health and environmental risks among risk assessors, risk managers, the public, the media, interested groups, and others.

Risk Management: The process of determining appropriate actions in response to an identified risk.

Risk Manager:
An individual, team, or organization with responsibility for or authority to take action in response to an identified risk.

Risk-Based Analysis: A process to help the risk and/or environmental manager better implement, direct, and use risk assessment and risk communication in order to influence the multi-component decisions involved in the risk management process.

Risk-Based Corrective Action: A streamlined approach in which exposure and risk assessment practices are integrated with traditional components of the corrective action process to ensure that appropriate and cost-effective remedies are selected, and that limited resources are properly allocated.

Safety: The probability that harm or loss will not occur under specified conditions.

Sediment: Particulate material lying below water.

Significant Risk: A risk that is an observed or measured event exceeding a defined threshold.

Site Use Factor: The ratio of an organism's home range, breeding range, or feeding/foraging range to the area of contamination of the site under investigation.

Situational Analysis: Identify, understand, and integrate the needs and objectives of others within the regulatory, political, and socioeconomic aspects of the property and their roles in risk management decision-making.

Stakeholder: Any individual, team, or organization interested in or affected by the outcome of a risk assessment.

Stressor: Any physical, chemical, or biological entity that can induce an adverse response.

Sustainable Development: A condition in which the environment or a component of the environment is renewed at essentially the same rate as it is used.

Toxicity: The inherent potential or capacity of a material to cause adverse effects in a living organism.

Toxicity Reference Value (TRV): A concentration above which some effect (or response) will be produced and below which it will not.

Trophic Level: A functional classification of taxa within a community that is based on feeding relationships (e.g., aquatic and terrestrial plants make up the first trophic level, and herbivores make up the second).

Uncertainty: A lack of confidence in the prediction of a risk assessment that may result from natural variability in natural processes, imperfect or incomplete knowledge, or errors in conducting an assessment.

Most terms and references cited in this glossary are from the USEPA's *Terms of Environment* (Revised, Document #EPA175B97001, 1997) available on the INTERNET at: http://www.epa.gov/reg5oopa/students/terms_of_environment.htm. Others come from select references cited throughout the text.

REFERENCES

Ackerman, J. and Soler, S.M., Upsizing Brownfield Sites: Creating Value Beyond the Surface, *Brownfields 2000 Conference Proceedings*, Track 4, Engineers' Society of W. Pennsylvania, Pittsburgh, PA, 7p, 2000.

Ackerman, J., Feinstein, J.L., and Roache, W.J., Environmental and Economic Fusion: The Brownfields Revitalization Process, *SiteWorks,* Report 101, Vanasse Hangen Brustlin (VHB), Inc., Watertown, MA, 52p, 1998.

Agency for Toxic Substances and Disease Registry (ATSDR), ATSDR's Toxicological Profiles on CD-ROM, CRCnetBASE, 1999.

Alberti, M., Modeling the Urban Ecosystem: A Conceptual Framework, *Environment and Planning B: Planning and Design* 26, 605 – 630, 1999.

Amdur, M.O., Doull, J., and Klaassen, C.D., *Casarett and Doull's Toxicology: The Basic Science of Poisons*, Fourth Edition, Pergamon Press, Inc., Elmsford, NY, 1006p, 1991.

American Society for Testing and Materials (ASTM), *Standard Guide for Risk-Based Corrective Action Applied at Petroleum Release Sites*, ASTM Designation E1739-95e1, American Society for Testing and Materials, West Conshohocken, PA, 1995.

American Society for Testing and Materials (ASTM), *Standard Provisional Guide for Risk-Based Corrective Action*, ASTM Designation PS 104-98, American Society for Testing and Materials, West Conshohocken, PA, 1998.

American Society for Testing and Materials (ASTM), *Standard Guide for Property Condition Assessments: Baseline Property Condition Assessment Process*, ASTM Designation E2018-99, American Society for Testing and Materials, West Conshohocken, PA, 1999.

American Society for Testing and Materials (ASTM), *Standard Guide for Environmental Site Assessments: Phase I Environmental Site Assessment Process*, ASTM Designation E1527-00, American Society for Testing and Materials, West Conshohocken, PA, 2000.

Ames, B.N. and Gold, L., Paracelsus to Parascience: The Environmental Cancer Distraction, *Mutation Research*, 447(1), 3 - 13, 2000.

Arnold, J.D., *The Complete Problem Solver*, J. Wiley & Sons, New York, 240p, 1992.

Barnard, R.C., Scientific Method and Risk Assessment, *Regulatory Toxicology and Pharmacology*, 19, 211 - 218, 1994.

Bélanger, P.J.P and Craig, A.J., Amortizing Risk Over Time: A New Approach to Brownfields Redevelopment, *SiteWorks* 2(5) pages 1 and 8, Vanasse Hangen Brustlin, Inc., Watertown, MA, July, 1999.

Bell, R., *Real Estate Damages—An Analysis of Detrimental Conditions*, Appraisal Institute, Chicago, IL, 361p, 1999.

Bradford-Hill, A., The Environment and Disease: Association or Causation, *Proc. Royal Soc. Med.*, 58, 295 - 300, 1966.

Bureau of National Affairs (BNA), *Environmental Due Diligence Guide*, Bureau of National Affairs, Washington, D.C., 2001.

Canter, L.W., *Environmental Impact Assessment*, Second Edition, McGraw Hill, Inc., New York, 660p, 1996.

Cantlon, J.E. and Koenig, H.E., Sustainable Ecological Economies, *Ecological Economics*, 31, 107 - 121, 1999.

Carpenter, R.A., Communicating Environmental Science Uncertainties, *Environmental Professional*, 17, 127 - 136, 1995.

Carroll, A.B., *Business and Society: Ethics and Stakeholder Management*, Second Edition, South-Western, Cincinnati, OH, 785p, 1993.

Chan Kim, W. and Mauborgne, R., Fair Process: Managing in the Knowledge Economy, *Harvard Business Review* 75(4), 65 - 75, July-August 1997.

Charnley, G., *Democratic Science: Enhancing the Role of Science in Stakeholder-Based Risk Management Decision-Making*, http://www.riskworld.com/Nreports/ 2000/Charnley/NR00GC00.htm, INTERNET, 2000.

Chechile, R.A. and Carlisle, S., *Environmental Decision Making: A Multi-disciplinary Perspective*, Van Nostrand Reinhold, New York, 296p, 1991.

Chechile, R.A., Chapter 1. Introduction to Environmental Decision Making, in *Environmental Decision Making: A Multi-disciplinary Perspective*, Van Nostrand Reinhold, New York, pp 1 – 13, 1991.

Chess, C., Organizational Theory and the Stages of Risk Communication, *Risk Analysis* 21(1), 179 – 188, 2001.

Conces, Rory J., *Blurred Visions—Philosophy, Science, and Ideology in a Trouble World*, Peter Lang, New York, 314p, 1997.

Conklin, A.R., What is Clean Soil, *Soil & Groundwater Cleanup*, 51-54, March 1996.

Costanza, R. and Cornwell, L., The 4P Approach to Dealing with Scientific Uncertainty, *Environment*, 34(9), 10p, http://dieoff.org/page33.htm, 1992.

Covey, S.R., *The 7 Habits of Highly Effective People*, Fireside, New York, NY, 358p, 1989.

Doty, C.B. and Travis, C.C., The Superfund Remedial Action Decision Process: A Review of Fifty Records of Decision, *J. Air Pollution Control Assoc.* 39(12), 1535 – 1543, 1989.

Earl, G. and Clift, R., Stakeholder Value Analysis: A Methodology for Integrating Stakeholder Values into Corporate Environmental Investment Decisions, *Business Strategy Environment* 8, 149 - 162, 1999.

Ecology and Environment, Inc., *Assessment and Evaluation of Risks to the Ecology and Public Health at the DDT Contamination Site, Bandelier National Monument, Los Alamos, New Mexico,* Ecology & Environment, Inc., Lancaster, NY, 50p, 1996.

Eisenhardt, K.M., Kahwajy, J.L., and Bourgeois, L.J., III, How Management Teams Can Have a Good Fight, *Harvard Business Review* 75, 77 - 85, 1997.

English, M.R., Feldman, D.L., Inerfeld, R., and Lumley, J., *Institutional Controls at Superfund Sites: A Preliminary Assessment of Their Efficacy and Public Acceptability,* Office of Policy, Planning, and Evaluation at the University of Tennessee, Knoxville, TN, 113p, 1997.

Environmental Law Institute (ELI), *A Guidebook for Brownfield Property Owners,* http://www.eli.org/pdf/rrguidebook99.pdf, INTERNET, 1999.

Ferguson, C.C., Assessing Risks from Contaminated Sites: Policy and Practice in 16 European Countries, *Land Contamination & Reclamation* 7(2), 87 - 107, 1999.

Fischhoff, B., Risk Perception and Communication Unplugged: Twenty Years of Process, *Risk Analysis* 21(1), 179 – 188, 1995.

Frantzen, K.A., Using Risk Appraisals to Manage Environmentally Impaired Properties, *SiteWorks,* Report 108, Vanasse Hangen Brustlin (VHB), Inc., Watertown, MA, 28p, 2000.

Freudenburg, W.R., Perceived Risk, Real Risk: Social Science and the Art of Probabilistic Risk Assessment, *Science* 242, 44 - 49, 1988.

Foster, K.R., Vecchia, P., and Repacholi, M.H., Science and the Precautionary Principle, *Science* 288, 979-980, 12 May 2000.

Gorczynski, D.M., *Insider's Guide to Environmental Negotiations,* Lewis Publishers/CRC Press, Inc., Boca Raton, FL, 242p, 1992.

Graham, J.D., Perspectives on the Precautionary Principle, *Human Environmental Risk Assessment* 6(3), 383 - 385, 2000.

Grose, V.L., *Managing Risk: Systematic Loss Prevention for Executives,* Omega Systems Group, Arlington, VA, 404p, 1987.

Hansen, M.T., Nohria, N., and Teirney, T., What's Your Strategy for Managing Knowledge, *Harvard Business Review* 77(2), 106 - 116, March-April 1999.

Hardin, G., The Tragedy of the Commons, *Science* 168, 1243 - 1248, 1968.

Harper, B. and Harris, S., *Using Eco-Cultural Risk in Risk-Based Decision Making,* International Institute for Indigenous Resource Management, Denver, CO, available from http://www.iiirm.org/publications/risk/risk.htm, INTERNET, 9p, 2001.

Harris, S. and Harper, B., *Environmental Justice in Indian Country: Using Equity Assessments to Evaluate Impacts to Trust Resources, Watersheds, and Eco-Cultural Landscapes,* International Institute for Indigenous Resource Management, Denver, CO, http://www.iiirm.org/publications/risk/risk.htm, INTERNET, 13p, 2001.

Harte, J., *Consider A Spherical Cow: A Course In Environmental Problem Solving,* University Science Books, Mill Valley, CA, 283p, 1988.

Havel, V., The Need for Transcendence in the Postmodern World, *Exploring Your Future,* World Future Society, Bethesda, MD, 141 – 144, 1996.

Head, G.L. and Horn, S., II, *Essentials of Risk Management,* Insurance Institute of America, Second Edition, Malvern, PA, 1991.

Hileman, B., Precautionary Principle, *Chemical and Engineering News* 76(6), 16 - 18, 9 February 1998.

Hoddinott, K.B. and Lee, A.P., The Use of Environmental Risk Assessment Methodologies for an Indoor Air Quality Investigation, *Chemosphere* 41, 77 - 84, 2000.

Houghton Mifflin, *The American Heritage Dictionary of the English Language,* Fourth Edition, Houghton Mifflin Company, Inc., Boston, MA, 2074p, 2000.

Huber P.W., Little Risks and Big Fears, *Regulatory Toxicology and Pharmacology* 7, 200 - 205, 1987.

Huning, A., *Preferences and Value Assessments in Cases of Decision Under Risk, Society for Phil. & Technol.* 4(4), 5p, http://scholar.lib.vt.edu/ejournals/SPT/v4n4/huning.html, INTERNET, 2000.

International Agency for Research on Cancer (IARC), *Overall Evaluations of Carcinogenicity to Humans,* http://193.51.164.11/monoeval/crthgr02a.html, INTERNET, 2000.

International Environmental Technology Centre (IETEC), *Environmental Risk Assessment for Sustainable Cities,* United Nations Environmental Programme, Technical Publication Series, Issue 3, Osaka/Shiga, Japan, 57p, 1996.

Jain, R.K., Urban, L.V., Stacey, G.S., and Balbach, H.E., *Environmental Assessment,* McGraw-Hill, Inc., New York, NY, 524p, 1993.

Kepner, C.H. and Tregoe, B.B., *The New Rational Manager,* Princeton Research Press, Princeton, NJ, 220p, 1981.

Kervern, G-Y., Cindynics: the Science of Danger, *Risk Management,* pp 34 – 42, March 1995.

Klaassen, C.D. and Eaton, D.L., Principles of Toxicology, in *Casarett and Doull's Toxicology, The Basic Science of Poisons,* Fourth Edition, Pergamon Press, Inc., Elmsford, NY, 2006p, 1991.

LaGoy, P.K. and Hopkins, L., Developing Site-Specific Cleanup Levels: Practical Considerations, *Remediation* 113 – 121, Spring, 1991.

Lehr, J.H., Toxicological Risk Assessment Distortions, The American Groundwater Trust, Water Well Journal Publishers, Dublin, OH, *Groundwater* 28(1), 2 - 8, 1990.

Liebs, L.H., Frantzen, K.A., and Ostrow, S.D., *Reality-Based Management of Environmentally-Impaired Properties: Lessons From Former Manufactured Gas Plant Sites (Balancing Societal, Regulatory, Financial, and Environmental Issues),* GEI Consultants, Inc./Atlantic Environmental Division, Colchester, CT, 49p, 1998.

Llewellyn, G., Strategic Risk Assessment—Prioritising Environmental Protection, *J. Hazardous Materials* 61, 279 - 286, 1998.

Lopez Cerezo, J.A., Lay Knowledge and Public Participation in Technological and Environmental Policy, *Society for Phil. & Technol.* 2(1), 16p, http://scholar.lib.vt.edu/ejournals/SPT/v2n1/cerezo.html, INTERNET, 1999.

Lubchenco, J., Entering the Century of the Environment: A New Social Contract for Science, *Science* 279, 291 - 497, 23 January 1998.

Madsen, H. and Ulhoi, J.P., Integrating Environmental and Stakeholder Management, *Business Strategy and the Environment* 10, 77 – 88, 2001.

McElroy, A. and Townsend, P.K., *Medical Anthropology in Ecological Perspective,* Third Edition, Westview Press, Boulder, CO, 434p, 1996.

Menzie, C. A., Risk Communication and Careful Listening—Resolving Alternative World Views, Introduction, *Human Ecol. Risk Assessment* 4(3), 619 – 622, 1998.

Morgan, M.G., Fischhoff, B., Bostrom, A., Lave, L., and Atman, C.J., Communicating Risk to the Public, Environ. Sci. Technol. 26(11), 2048 – 2056, 1992.

Murphy, K.G. and Fitzgerald, E., Jr., Risk Analysis—Working With the Customer, *Risk Management Quarterly* 2(2), 1, 1994.

National Research Council (NRC), *Risk Assessment in the Federal Government: Managing the Process,* National Academy Press, Washington, D.C., 189p, 1983.

National Research Council (NRC), *Science and Judgment in Risk Assessment,* National Academy Press, Washington, D.C., 651p, 1994.

National Research Council (NRC), *Understanding Risk: Informing Decisions in a Democratic Society,* National Academy Press, Washington, D.C., 249p, 1996.

Neuman, S., *Integrated Environmental Risk Management in Real Estate Transactions,* available via http://www.environmental-center.com/articles/article 734/article734.htm, INTERNET, 1998.

Pope, R.H., *PRA—A Risk Manager's View,* presented at the Third Workshop on Practical Issues in the Use of Probabilistic Risk Assessment – PRA Step-by-Step, Safety Harbor Resort, Tampa, FL, 16-18April 2000.

Portney, K.E., Chapter 9. Public Environmental Policy Decision Making: Citizen Roles, in *Environmental Decision Making: A Multi-Disciplinary Perspective,* Van Nostrand Reinhold, New York, pp 195 – 216, 1991.

Presidential/Congressional Commission on Risk Assessment and Risk Management (PCCRARM), *Framework for Environmental Health Risk Management,* Final Report, Volume 1, PCCRARM, Washington, D.C., 64p, http://www.riskworld.com, INTERNET, 1997a.

Presidential/Congressional Commission on Risk Assessment and Risk Management (PCCRARM), *Risk Assessment and Risk Management in Regulatory Decision-Making,* Final Report, Volume 2, PCCRARM, Washington, D.C., 213p, http://www.riskworld.com, INTERNET, 1997b.

Presidential/Congressional Commission on Risk Assessment and Risk Management (PCCRARM), *A Public Health Approach to Environmental Protection,* PCCRARM, Washington, D.C., 20p, http://www.riskworld.com, INTERNET, 1997c.

Price, P.P. and Keenan, R., Probabilistic Techniques in Risk Assessment: Make Room in the Regulatory Toolbox, *Perspectives on Risk Policy: Behind the Bitter Debate,* Risk Policy Report, Inside Washington Publishers, Washington, D.C., pp 109 – 111, 1997.

Rein, M., Value–Critical Policy Analysis, In *Ethics, The Social Sciences and Policy Analysis,* Ed. D. Callahan and B. Jennings, Kluwer Academic/Plenum Publishers, New York, 408p, 1983.

Rejda, G. E., *Principles of Risk Management and Insurance,* Fourth Edition, Harper Collins Publishers, New York, 773p, 1992.

Renn, O., A Model for an Analytic-Deliberative Process in Risk Management, *Environ. Sci. & Technol.* 33(18), 3049 - 3055, 1999.

Rio Convention, United Nations Conference on Environment and Development: Rio Declaration on Environment and Development, June 14, 1992, as printed in *International Legal Materials* 31, 874 – 879, 1992.

Risk Assessment Information Service (RAIS), available from http://risk.lsd.ornl.gov/rap_hp.htm, INTERNET, 2000.

Robinson, A., ed., *Continuous Improvement in Operations: A Systematic Approach to Waste Reduction,* Productivity Press, Cambridge, MA, 406p, 1993.

Robinson, S., Key Survival Issues: Steps Toward Corporate Environmental Sustainability, *Corporate Environmental Strategy* 7(1), 92 - 105, 2000.

Roome, N., Conceptualizing and Studying the Contribution of Networks in Environmental Management and Sustainable Development, *Business Strategy and the Environment* 10, 69 - 76, 2001.

Ross, J.F., Risk: Where Do Real Dangers Lie, *Smithsonian* 26(8), 43 – 53, 1995.

Rugen, P. and Callahan, B., An Overview of Monte Carlo: A Fifty Year Perspective, *Human Ecological Risk Assessment* 2(4), 671 - 680, 1996.

Sample, B.E., Opresko, D.M., and Suter, G.W., Toxicological Benchmarks for Wildlife: 1996 Revision, ES/ER/TM-86/R3, Prepared for the U.S. Department of Energy, Office of Environmental Management, Oak Ridge National Laboratory, Oak Ridge, TN, 250p, 1996.

Sandin, P., Dimensions of the Precautionary Principle, *Human Ecological Risk Assessment* 5, 889 - 907, 1999.

Schierow, L., *Risk Analysis and Cost-Benefit Analysis of Environmental Regulations*, Congressional Research Service Report for Congress, Environment and Natural Resources Policy Division, Washington, D.C., http://www.cnie. org/nle, INTERNET, 1994.

Schrader-Frechette, K.S., *Risk and Rationality*, University of California Press, Berkeley, CA, 312p, 1991.

Schulz, T.W. and Griffin, S., Practical Methods for Meeting Remediation Goals at Hazardous Waste Sites, *Risk Analysis* 21(1), 43 - 52, 2001.

Slovic, P., Perception of Risk, *Science* 236, 280 – 285, 1987.

Smith, G.F., *Quality Problem Solving*, ASQ Quality Press, Milwaukee, WI, 322p, 1998.

Stern, A., The Future of Monte Carlo Analysis in Human Health Risk Assessment, *Risk Policy Report* 4(11), 37-40, 1997.

Strasser, A., Expediting Brownfields through Negotiation: A Best Practices Primer, *Brownfield News* 38 - 40, May/June 2000.

Susskind, L. and Field, P., *Dealing with an Angry Public: The Mutual Gains Approach to Resolving Disputes*, The Free Press, New York, 276p, 1996.

Suter, II, G.W., Developing Conceptual Models for Complex Ecological Risk Assessments, *Human Ecol. Risk Assess.* 5(2) 375 – 396, 1999.

Swaffield, S., Frames of Reference: A Metaphor for Analyzing and Interpreting Attitudes of Environmental Policy Makers and Policy Influencers, *Environmental Management* 22(4), 495 - 504, 1998.

Tal, A., Assessing the Environmental Movement's Attitudes Toward Risk Assessment, *Environ. Sci. & Technol.* 31(10), 470A - 476A, 1997.

Tillich, P., *A History of Christian Thought*, Braaten, C.E., Ed., Touchstone/Simon & Schuster, Inc., New York, 550p, 1968.

Tillich, P., *Systematic Theology*, Volume I, The University of Chicago Press, Chicago, IL, 289p, in three volume set published 1967.

Tissembaum, C.A., So You Want to be a Project Engineer, *Chemical Engineering*, pp. 169–174, 1993.

Toll, J.E., Elements of Environmental Problem-Solving, *Human and Ecological Risk Assessment* 5(2), 275 - 280, 1999.

Tong, R., *Ethics in Policy Analysis*, Prentice-Hall, Inc., Englewood Cliffs, NJ, 200p, 1986.

United States Environmental Protection Agency (USEPA), Guidelines for Carcinogen Risk Assessment, *Federal Register* 51:33992-34012, Washington, D.C., 1986a.

United States Environmental Protection Agency (USEPA), Risk Assessment Guidelines for Mutagenicity Risk Assessment, *Federal Register* 51:34006, Washington, D.C., September 24, 1986b.

United States Environmental Protection Agency (USEPA), *Superfund Public Health Evaluation Manual*, EPA/540/1-86/060 (OSWER Directive 9285.4-1), Office of Emergency and Remedial Response, Washington, D.C., 1986c.

United States Environmental Protection Agency (USEPA), *Guidance for Conducting Remedial Investigations and Feasibility Studies Under CERCLA*, EPA/540/G-89/004 (OSWER Directive 9355.3-01), Office of Emergency and Remedial Response, Washington, D.C., 1988.

United States Environmental Protection Agency (USEPA), *Exposure Factors Handbook*, EPA/600/8-89/043, Office of Research and Development, Washington, D.C., 1989a.

United States Environmental Protection Agency (USEPA), *Risk Assessment Guidance for Superfund (RAGS): Volume I - Human Health Evaluation Manual (HHEM) (Part A, Baseline Risk Assessment)*, Interim Final, EPA/540/1-89/002, NTIS PB90-155581, Office of Emergency and Remedial Response, Washington, D.C., 1989b.

United States Environmental Protection Agency (USEPA), *Risk Assessment Guidance for Superfund, Volume II, Environmental Evaluation Manual*, Interim Final, EPA/540/1-89/001A (OSWER Directive 9285.7-01), Office of Emergency and Remedial Response, Washington, D.C., 1989c.

United States Environmental Protection Agency (USEPA), *Interim Final Guidance on Preparing Superfund Decision Documents*, Office of Emergency and Remedial Response, OSWER Directive 9355.3-02, USEPA, Washington, D.C., 1989d.

United States Environmental Protection Agency (USEPA), *A Guide to Selecting Superfund Remedial Actions*, OSWER Directive 9355.0-27FS, Office of Solid Waste and Emergency Response, Washington, D.C., 1990a.

United States Environmental Protection Agency (USEPA), *Guidance for Data Usability in Risk Assessment (Part A)*, Final, OSWER Directive 9285.7-09A, PB92-963356, USEPA, Office of Emergency and Remedial Response, Washington, D.C., 1990b.

United States Environmental Protection Agency (USEPA), *Guidance for Data Usability in Risk Assessment (Part B)*, Final, OSWER Directive 9285.7-09B, PB92-963362, Office of Emergency and Remedial Response, Washington, D.C., 1990c.

United States Environmental Protection Agency (USEPA), National Oil and Hazardous Substance Pollution Contingency Plan, 40 CFR 300, 55 *Federal Register*: 8666, Washington, D.C., Thursday 8 March 1990d.

United States Environmental Protection Agency (USEPA), *Update on OSWER Soil Lead Cleanup Guidance*, OSWER Directive 9355.4-02a, Office of Emergency and Remedial Response, Washington, D.C., August 29, 1991a.

United States Environmental Protection Agency (USEPA), *Risk Assessment Guidance for Superfund (RAGS): Volume I - Human Health Evaluation Manual (HHEM) (Part B, Development of Risk-Based Preliminary Remediation Goals)*, EPA/540/R-92/003, OSWER Directive 9285.7-01B, NTIS PB92-963333, Office of Emergency and Remedial Response, Washington, D.C., 1991b.

United States Environmental Protection Agency (USEPA), *Risk Assessment Guidance for Superfund (RAGS) Volume I: Human Health Evaluation Manual (HHEM) (Part C, Risk Evaluation of Remedial Alternatives)*, Interim, EPA/540/R-92/004, OSWER Directive 9285.7-01C, NTIS PB92-963334, Office of Emergency and Remedial Response, Washington, D.C., 1991c.

United States Environmental Protection Agency (USEPA), *Risk Assessment Guidance for Superfund (RAGS): Volume I - Human Health Evaluation Manual Supplemental Guidance: Standard Default Exposure Factors*, Interim Final, OSWER Directive 9285.6-03, Office of Emergency and Remedial Response, Washington, D.C., 1991d.

United States Environmental Protection Agency (USEPA), *Guidance on Oversight of Potentially Responsible Party Remedial Investigations and Feasibility Studies*, Final, OSWER Directive No. 9835.1(c), Office of Solid Waste and Emergency Response, Washington, D.C., 1991e.

United States Environmental Protection Agency (USEPA), *Role of the Baseline Risk Assessment in Superfund Remedy Selection Decisions*, OSWER Directive 9355.0-30, Office of Solid Waste and Emergency Response, Washington, D.C., 1991f.

United States Environmental Protection Agency (USEPA), *Framework for Ecological Risk Assessment*, EPA/630/R-92/001, Office of Solid Waste and Emergency Response, Washington, D.C., 1992a.

United States Environmental Protection Agency (USEPA), *Data Quality Objectives Process for Superfund*, Interim Final Guidance, OSWER Directive 9355.9-01, EPA/540/R-93/071, Office of Solid Waste and Emergency Response, Washington, D.C., 1992b.

United States Environmental Protection Agency (USEPA), *Dermal Exposure Assessment: Principles and Applications*, EPA/600/8-91/011B, Office of Health and Environmental Assessment, Washington, D.C., 1992c.

United States Environmental Protection Agency (USEPA), *Human Health Evaluation Manual: Supplemental Guidance: Interim Dermal Risk Assessment Guidance*, OSWER Directive 9285.7-10, Office of Solid Waste and Emergency Response, Washington, D.C., 1992d.

United States Environmental Protection Agency (USEPA), *Final Guidance on Data Usability in Risk Assessment (Part A)*, OSWER Directive 9285.7-09A, Office of Solid Waste and Emergency Response, Washington, D.C., 1992e.

United States Environmental Protection Agency (USEPA), *Final Guidance on Data Usability in Risk Assessment (Part B)*, Office of Solid Waste and Emergency Response, OSWER Directive 9285.7-09B, USEPA, Washington, D.C., 1992f.

United States Environmental Protection Agency (USEPA), *Supplemental Guidance to RAGS: Calculating the Concentration Term*, OSWER Directive 9285.7-081, Office of Solid Waste and Emergency Response, Washington, D.C., 1992g.

United States Environmental Protection Agency (USEPA), *Guidance for Conducting Non-Time Critical Removal Actions Under CERCLA*, EPA/540/R-93/057, Office of Solid Waste and Emergency Response, Washington, D.C., 1993a.

United States Environmental Protection Agency (USEPA), *Provisional Guidance for Quantitative Risk Assessment of Polycyclic Aromatic Hydrocarbons*, Office of Research and Development, EPA/600/R-93/C89, USEPA, Washington, D.C., 1993b.

United States Environmental Protection Agency (USEPA), *Revised Interim Soil Lead Guidance for CERCLA Sites and RCRA Corrective Action Facilities*, OSWER Directive 9355.4-12, Office of Solid Waste and Emergency Response, Washington, D.C., 1993c.

United States Environmental Protection Agency (USEPA), Data Quality Objectives Process for Superfund: Interim Final Guidance, EPA/540/G-93/071, Washington, D.C., 1994.

United States Environmental Protection Agency (USEPA), EPA Risk Characterization Program, Memorandum from Administrator Carol Browner, Office of the Administrator, Washington, D.C., March 21, 1995a.

United States Environmental Protection Agency (USEPA), *Policy for Risk Characterization*, http://www.epa.gov/nceawww1/riskchar.htm, INTERNET, 1995b.

United States Environmental Protection Agency (USEPA), *Use of Monte Carlo Simulation in Performing Risk Assessments*, Region 8 Superfund Technical Guidance, Washington, D.C., 1995c.

United States Environmental Protection Agency (USEPA), Soil Screening Guidance: Technical Background Document, Office of Solid Waste and Emergency Response, EPA/540/R-95/126, USEPA, Washington, D.C., 1995d.

United States Environmental Protection Agency (USEPA), *Memorandum* from Carol Browner on Risk Characterization, Office of the Administrator, Washington, D.C., February 22, 1995d.

United States Environmental Protection Agency (USEPA), *Proposed Guidelines for Ecological Risk Assessment*, EPA/630/R-95/002B, Risk Assessment Forum, Washington, D.C., 1996a.

United States Environmental Protection Agency (USEPA), *Exposure Factors Handbook* - SAB Review Document, Office of Health and Environmental Assessment, Washington, D.C., 1996b.

United States Environmental Protection Agency (USEPA), *Final Soil Screening Guidance, Soil Screening Guidance User's Guide*, EPA/540/R-96/018, Office of Solid Waste and Emergency Response, Washington, D.C., May 17, 1996c.

United States Environmental Protection Agency (USEPA), *PCBs: Cancer Dose-Response Assessment and Application to Environmental Mixtures*, EPA/600/P-96/001A, Office of Research and Development, Washington, D.C., 1996d.

United States Environmental Protection Agency (USEPA), *Ecological Risk Assessment Guidance for Superfund: Process for Designing and Conducting Ecological Risk Assessments*, Interim Final, U.S. EPA Environmental Response Team, Edison, NJ, 1997a.

United States Environmental Protection Agency (USEPA), *EPA Guidance for Data Assessment*, EPA/600/R-96/084, Office of Research and Development, Washington, D.C., 1997b.

United States Environmental Protection Agency (USEPA), *Rules of Thumb for Superfund Remedy Selection*, OSWER 9355.0-69, EPA/540-R-97-013, Office of Solid Waste and Emergency Response, Washington, D.C., 1997c.

United States Environmental Protection Agency (USEPA), *Rules of Thumb for Superfund Remedy Selection*, OSWER 9355.0-69, EPA/540-R-97-013, Office of Solid Waste and Emergency Response, Washington, D.C., 1997d.

United States Environmental Protection Agency (USEPA), *Ecological Risk Assessment Guidance for Superfund: Process for Designing and Conducting Ecological Risk Assessments*, Office of Solid Waste and Emergency Response, OSWER 9285.7-25, EPA/540-R-97-006, USEPA, Washington, D.C., 1997e.

United States Environmental Protection Agency (USEPA), *Guiding Principles for Monte Carlo Analysis*, EPA/630/R-97/001, Office of Research and Development, Washington, D.C., 1997f.

United States Environmental Protection Agency (USEPA), *Health Effects Assessment Summary Tables (HEAST)*, FY 1997 Update, Office of Solid Waste and Emergency Response, EPA-540-R-97-036, USEPA, Washington, D.C., 1997g.

United States Environmental Protection Agency (USEPA), *Policy for Use of Probabilistic Analysis in Risk Assessment*, EPA/630/R-97/001, Office of Research and Development, Washington, D.C., 1997h.

United States Environmental Protection Agency (USEPA), *Exposure Factors Handbook*, Volumes I, II, and III, EPA/600/P-95/002Fa, NTIS: PB98-124217, Office of Research and Development, Washington, D.C., http://www.epa.gov/ncea, INTERNET, 1998a.

United States Environmental Protection Agency (USEPA), *Guidelines for Ecological Risk Assessment*, Final, EPA/630/R-95/002F, Risk Assessment Forum, Washington, D.C., 1998b.

United States Environmental Protection Agency (USEPA), *Federal Facilities Stakeholder Involvement: Blueprint for Action*, Federal Facilities Restoration and Reuse, EPA505-F-99-006, Office Solid Waste and Emergency Response, Washington, D.C., 1999a.

United States Environmental Protection Agency (USEPA), *Risk Assessment Guidance for Superfund: Volume 3 (Part A, Process for Conducting Probabilistic Risk Assessment)*, Draft, Revision No. 5, Office of Solid Waste and Emergency Response, Washington, D.C., 1999b.

United States Environmental Protection Agency (USEPA), *Glossary of Integrated Risk Information System Terms*, Washington, D.C., http://www.epa.gov/ngispgm3/iris/gloss8.htm, INTERNET, 1999c.

United States Environmental Protection Agency (USEPA), *Data Quality Objectives Process for Hazardous Waste Site Investigations*, Final, EPA QA/G-4HW, EPA/600-R-00/007, Office of Environmental Information, Washington, D.C., 2000.

United States Environmental Protection Agency (USEPA), National Contingency Plan, *Code of Federal Regulation* Title 40 Part 300, http://www.access.gpo.gov/nara/cfr/waisidx_00/40cfr300_00.html, INTERNET, 2001a.

United States Environmental Protection Agency (USEPA), *Integrated Risk Information System (IRIS)*, http://www.epa.gov/ngispgm3/iris/subst/index.html, INTERNET, 2001b.

Vanderzwaag, Z., *CEPA and the Precautionary Principle*, Canadian Environmental Protection Agency, Ottawa, Canada, http://www.ec.gc.ca/cepa/ip18/e18_00.html, INTERNET, 1999.

Vincent, R., *The "R" Word: Risk Assessment in Perspective*, Sierra Club, Pueblo, CO, 6p, 1999.

Voorhees, J. and Woellner, R.A., *International Environmental Risk Management: ISO 14000 and the Systems Approach*, Lewis Publishers, CRC Press LLC, Boca Raton, FL, 268p, 1998.

Walker, K.D., Sadowitz, M., and Graham, J.D., *Confronting Superfund Mythology: The Case of Risk Assessment and Risk Management*, Center for Risk Analysis, Harvard School of Public Health, Boston, MA, 17p, 1994.

Walker, V.R., Risk Characterization and the Weight of Evidence: Adapting Gatekeeping Concepts from the Courts, *Risk Analysis* 16(6), 793 - 799, 1996.

Washburn, S.T. and Edelmann, K.G., Development of Risk-Based Remediation Strategies, *Practice Periodical of Hazardous, Toxic, and Radioactive Waste Management* 3(2), 77 – 82, 1999.

Williams, C.A., Freeman, R.W., and James, R.C., Availability and Use of Human Data in the Development of USEPA Reference Doses and Reference Concentrations, *The Toxicologist* 15(1), 33 – 34, 1995.

Wilson, A.R., *Environmental Risk: Identification and Management*, Lewis Publishers, CRC Press LLC, Boca Raton, FL, 285p, 1991.

Wolfe, A.M., Risk Communication in Social Context: Improving Effective Communication, *Environ. Professional* 15, 248 – 255, 1993.

Wynne, B., The Institutional Context of Science, Models, and Policy: The IIASA Energy Study, *Policy Sciences* 17, 277 – 320, 1984.

Young, F.E., Risk Assessment: The Convergence of Science and the Law, *Reg. Toxicol. & Pharmacol.* 7, 179 – 184, 1987.

Zagaski, Jr., C.A., *Environmental Risk and Insurance*, Lewis Publishers, CRC Press LLC, Boca Raton, FL, 652p, 1992.

INDEX

A

Abiotic systems, risk system residing within, 6
Acrolein, 65
Active remediation, 131
Activity limitation, 116
Actual risk, 21, 22
Agency sign-offs, 123
Agents of change, 1
Air pollutants, carcinogenic potency of hazardous, 133
Alternative analysis, 93, 119
American Society for Testing and Materials (ASTM), 9, 32
    RBCA Tier 1 screening levels, 77
    standard, chemical release sites, 132
Analytic-deliberative process, 136
Applicable or relevant and appropriate requirements (ARARs), 81
    chemical-specific, 143, 199
    types of, 81
ARARs, see Applicable or relevant and appropriate requirements
Area infrastructure, 120
Assessment
    endpoints, 65, 70, 99
    /management/communication paradigm, 59
Asset definition, 93, 119
ASTM, see American Society for Testing and Materials
Attorney-client privilege, 167
Austrian Water Act, 155
Avoidance tendency, 16

B

Background
    exposures, health risks associated with, 79
    information, assembling of available, 99
    knowledge, 41
Bandelier National Monument, 49
Baseline risks, 69
Bhopal, 86
Biodiversity, impact of company products on, 30
Biological agents, 188
Biotic systems, risk system residing within, 6
Birth defects, 64
Brownfields
    remediation project, 162
    revitalization process, 14

    sites, 12
Budgets, metrics for meeting of annual, 160
Building contamination, 29
Bureaucratic processes, performance-driven processes versus, 123
Businesses, as action-takers, 140
Buy or no-buy decisions, 15

C

CAA, see Clean Air Act
Cancer, 48, 75
    bioassay studies, 63
    risk(s), 77, 155
        assessments, 31
        incremental lifetime, 50
        lifetime excess, 67
        total excess, 68
    slope factors (CSF), 67
Capital expenditures, 108, 160, 161, 162
Capital gains, 28, 166
CARACAS, see Concerted Action on Risk Assessment for Contaminated Sites
Carcinogen(s)
    individual risks for, 67
    meaning of, 67
    preliminary remediation goals for, 199
    risks, 68, 84
    slope factors, 150
    substances, 62, 63
Carson, Rachel, 19
Catastrophes, solution to, 30
Cause-and-effect, 3
CBEP program, see USEPA Community Based Environmental Protection program
CERCLA, see Comprehensive Environmental Response, Compensation, and Liability Act
Chemical(s)
    armamentarium, diverse human needs served by, 20
    arts, liabilities associated with use of, 20
    constituents, 188
    contamination, 22, 38
    frequency of detection of, 72
    release sites, ASTM standard for, 132
    sources of, 66
    stressors, 101
    volatile, 157
Children's playgrounds, risks assess for, 155